Genotoxicity and Carcinogenicity Testing of Pharmaceuticals

Michael J. Graziano • David Jacobson-Kram
Editors

Genotoxicity and Carcinogenicity Testing of Pharmaceuticals

Springer

Editors
Michael J. Graziano
Vice- President of Drug Safety Evaluation
Bristol-Myers Squibb
Princeton
New Jersey
USA

David Jacobson-Kram
ToxRox Consulting
McLean
Virginia
USA

ISBN 978-3-319-22083-3 ISBN 978-3-319-22084-0 (eBook)
DOI 10.1007/978-3-319-22084-0

Library of Congress Control Number: 2015954554

Springer Cham Heidelberg New York Dordrecht London

Printed on acid-free paper

Springer International Publishing AG Switzerland is part of Springer Science+Business Media (www.springer.com)

Editors

Mike Graziano has more than 27 years of experience as a toxicologist in the pharmaceutical industry. He is currently Vice President of Drug Safety Evaluation at Bristol-Myers Squibb where he is responsible for the nonclinical safety testing for all drugs within the BMS portfolio. Within the past 6 years, Mike has provided nonclinical support for eight new drug approvals. Prior to joining BMS in 2003, he was Director of Anticancer and Antibacterial Toxicology Programs at Pfizer Pharmaceutical Research, Ann Arbor, MI, and with Parke-Davis Pharmaceutical Research Division/Warner-Lambert Company. He is author/co-author of more than 90 scientific publications and abstracts, many dealing with the nonclinical safety of new therapeutic agents. Mike is a member of the IQ DruSafe Leadership Group, the PhRMA Clinical and Preclinical Development Committee, the EFPIA Preclinical Safety Development Committee, and HESI. He is also the PhRMA Deputy Topic Leader for the current ICH S1 Expert Working Group. Mike received a B.S. in Animal Science from Rutgers University, a M.S. in Veterinary Toxicology from Louisiana State University, a Ph.D. in Toxicology from the University of Kentucky, and was a Postgraduate Research Toxicologist at the University of California, Berkeley. Mike is also a Diplomate of the American Board of Toxicology.

David Jacobson-Kram received his PhD in embryology from the University of Connecticut; he then went on to serve as a staff fellow at the National Institute on Aging. After leaving N.I.H., Dr. Jacobson-Kram joined the faculty of George Washington University School of Medicine and then later, Johns Hopkins University Oncology Center. During this same period he served, on a part-time basis, as a geneticist in the Office of Toxic Substances at the Environmental Protection Agency and as Acting Branch Chief in EPA's Office of Research and Development. Dr. Jacobson-Kram served as the VP of the Toxicology and Laboratory Animal Health Division at BioReliance Corporation, a contract testing laboratory from 1988 until 2003. From 2003 until 2014, he served as the Associate Director of Pharmacology and Toxicology in FDA's Office of New Drugs. Over the past thirty years, he has served as principal and co-principal investigator on several N.I.H. grants and government contracts and published widely in the areas of genetic and molecular toxicology. Dr. Jacobson-Kram has served as council member, treasurer, and chairman of the Genetic Toxicology Association; executive council member to the Environmental Mutagen Society; editor of *Cell Biology and Toxicology*; and a member of N.I.H. special study sections. In 1996, he became a Diplomate of the American Board of Toxicology (DABT).

Contents

Contributors

Jiri Aubrecht Drug Safety Research and Development, Pfizer Inc., Groton, CT, USA

Francesco Bellanti Division of Pharmacology, Leiden Academic Center for Drug Research, Leiden, The Netherlands

Todd Bourcier, PhD US Food and Drug Administration, Center for Drug Evaluation and Research, Silver Spring, MD, USA

Lorrene A. Buckley Lilly Research Laboratories, Toxicology and Drug Disposition, Eli Lilly and Company, Inc., Indianapolis, IN, USA

L.L. Custer Drug Safety Evaluation, Bristol-Myers Squibb, New Brunswick, NJ, USA

Maggie Dempster Safety Assessment, GlaxoSmithKline, LLC., Philadelphia, PA, USA

Stephen D. Dertinger Litron Laboratories, Rochester, NY, USA

Ruben G. Duijnhoven Section on Pharmacovigilance, Medicines Evaluation Board, Utrecht, The Netherlands

Division of Pharmacoepidemiology and Clinical Pharmacology, Utrecht Institute for Pharmaceutical Sciences, Utrecht University, Utrecht, The Netherlands

Annette Emerenciana Division of Pharmacology, Leiden Academic Center for Drug Research, Leiden, The Netherlands

Patricia A. Escobar Merck Research Laboratories, West Point, PA, USA

Michael J. Graziano Drug Safety Evaluation, Bristol-Myers Squibb, Princeton, NJ, USA

Robert H. Heflich Division of Genetic and Molecular Toxicology, National Center for Toxicological Research/U.S. Food and Drug Administration (U.S. FDA/NCTR), Jefferson, AR, USA

Abigail Jacobs US Food and Drug Administration, Center for Drug Evaluation and Research, Silver Spring, MD, USA

David Jacobson-Kram ToxRox Consulting, McLean, VA, USA

Peter Kasper Genetic and Reproductive Toxicology, Federal Institute for Drugs and Medical Devices (BfArM), Bonn, Germany

Naomi L. Kruhlak US Food and Drug Administration, Center for Drug Evaluation and Research, Silver Spring, MD, USA

Marcel Kwa Section on Pharmacovigilance, Medicines Evaluation Board, Utrecht, The Netherlands

James S. MacDonald Synergy Partners R&D Solutions, LLC, Chester, NJ, USA

Lutz Müller F. Hoffmann La-Roche, Pharmaceutical Sciences, Roche Innovation Center Basel, Basel, Switzerland

M.W. Powley U.S. Food and Drug Administration, Center for Drug Evaluation and Research, Silver Spring, MD, USA

Denis Roy Nonclinical Drug Safety, SciLucent LLC, Vista, CA, USA

Kris Siezen Section on Pharmacological, Toxicological and Biotechnological Assessment, Medicines Evaluation Board, Utrecht, The Netherlands

Beatriz Silva-Lima iMED.ULisboa, Faculdade de Farmacia, Universidade de Lisboa, Lisboa, Portugal

Frank D. Sistare Safety Assessment and Laboratory Animal Resources, Merck and Co., Inc., West Point, PA, USA

Lidiya Stavitskaya US Food and Drug Administration, Center for Drug Evaluation and Research, Silver Spring, MD, USA

Bas ter Braak Division of Toxicology, Leiden Academic Center for Drug Research, Leiden, The Netherlands

Mark A. Tirmenstein Drug Safety Evaluation, Bristol-Myers Squibb Company, New Brunswick, NJ, USA

Willem van den Brink Division of Pharmacology, Leiden Academic Center for Drug Research, Leiden, The Netherlands

Jan Willem van der Laan Section on Pharmacological, Toxicological and Biotechnological Assessment, Medicines Evaluation Board, Utrecht, The Netherlands

Division of Toxicology, Leiden Academic Center for Drug Research, Leiden, The Netherlands

Center for Health Protection, National Institute for Public Health and the Environment, Bilthoven, The Netherlands

Chapter 1
Introduction: An Overview of Industry and Regulatory Perspectives on the Genotoxic and Carcinogenic Assessment of Pharmaceuticals

Michael J. Graziano

Abstract While there are numerous manuscripts and review articles that cover various aspects of genotoxicity and carcinogenicity testing of pharmaceuticals, there is no single text book that brings all of these concepts together in a practical way. Therefore, the intent of this book is to help industry scientists and regulators develop a more comprehensive understanding of the concepts and strategies used to assess risk of these critical components of a nonclinical testing program. Assessing the risks for genotoxicity and carcinogenicity of pharmaceuticals differs from other chemicals since pharmaceuticals are given intentionally at relatively high doses in order to achieve a therapeutic benefit. Therefore, the safety assessment of pharmaceuticals, including genotoxicity and carcinogenicity evaluations, is often based on defining acceptable therapeutic margins and establishing the human relevance of findings in the animal studies. This book focuses on these topics in an integrated way, taking into account the rapid advances in safety sciences and evolving regulatory requirements. The book is written by well recognized experts from the pharmaceutical industry and US and European health authorities. All of the authors have either addressed various nonclinical safety issues over the course of their careers, were involved in developing the testing guidelines, and/or are thought leaders that continue to drive the science of toxicology forward. The order of the chapters reflects the usual sequence of genotoxicity and carcinogenicity testing in the pharmaceutical industry, starting with structure-based assessments very early in the drug development process. The book is also intended help readers better understand and appreciate the complexity of the regulations and breadth of toxicology research that are necessary to support the development of new drugs. Developing new drugs is extremely difficult as the expectations for safety continue to increase and target biology becomes more complex. These factors combined with the pressure to reduce animal use makes nonclinical safety testing challenging in today's environment. The last few years indicate that we are

M.J. Graziano
Drug Safety Evaluation, Bristol-Myers Squibb, Princeton, NJ, USA
e-mail: michael.graziano@bms.com

M.J. Graziano, D. Jacobson-Kram (eds.), *Genotoxicity and Carcinogenicity Testing of Pharmaceuticals*, DOI 10.1007/978-3-319-22084-0_1

at the cusp of major changes in the nonclinical safety testing of pharmaceuticals as evidenced by the number of new and revised ICH guidances along with advances in the development and application of in vitro safety assays. This book attempts to bring it all together as a "state of the science" and practical guide with references to numerous examples and important case studies. The Introduction provides a brief overview of each chapter and highlights some of the key considerations and approaches for de-risking drug development programs.

Keywords Carcinogenicity • Genotoxicity • ICH • Pharmaceuticals • Rodent bioassays

The realization that exposure to certain chemicals could lead to cancer originated well over a century ago based on observations of increased testicular cancer in chimney sweeps and increased urinary bladder cancer in workers in the dye industry. This was followed decades later by clear experimental evidence of chemical-induced tumors in animal studies following topical or oral administration of coal tar, polycyclic aromatic hydrocarbons, 2-naphthylamine, and azo dyes. A detailed history of chemical induced carcinogenesis can be found in numerous review articles and textbooks [1–5].

In response to the growing appreciation that some chemicals could lead to cancer and the need to protect public health, the US government enacted legislation in 1962 that required drug manufacturers to prove that their products were both safe and effective prior to marketing approval [6]. Around that time, the FDA also produced the first set of guidelines for preclinical safety testing [7]. Although these initial guidelines did not specifically state the need for carcinogenicity testing, chronic studies up to 18 months were recommended. Over the next few decades, protocols for carcinogenicity studies were refined and standardized. However, even today, they are largely based on the protocols developed by the NCI in the 1960s [7, 8]. Ironically, despite all of the technical and scientific innovations over the last 50 years, a 2-year study in rats is still considered the gold standard for carcinogenicity testing by regulatory authorities [9].

By the 1970s it was generally accepted that the mechanism of chemical-induced carcinogenesis involved interaction of the chemical with host DNA either by direct binding of the parent molecule or through the formation of reactive intermediates by the cytochrome P450 drug metabolizing enzyme system [1, 3, 10–12]. If not repaired, DNA binding of these reactive chemicals could lead to mutations in the genetic code and, ultimately, transformation of normal cells to cancer. Although this is an overly simplified description of chemical-induced carcinogenesis, it is evident that the early events in this process could be investigated without the use of animals. Accordingly, Bruce Ames et al. developed a relatively simple in vitro assay to detect chemical mutagens using *Salmonella typhimurium* bacteria and a mammalian drug metabolizing enzyme system [13]. It is now generally accepted that the "Ames assay," as it is commonly known, can detect DNA-reactive carcinogens with a fairly high degree of concordance [14–20]. Based on this high degree of sensitivity for identifying multi-site and multi-species carcinogens, the Ames assay is used within the pharmaceutical industry as an early screening assay and is part of the core battery of genotoxicity tests required by regulatory authorities [21, 22]. With some exceptions (e.g., cytotoxic anticancer drugs), almost all positives in this assay are dropped from development.

While the current pharmaceutical testing paradigm for genotoxicity and carcinogenicity testing generally works well, it has not changed dramatically for several decades and there is a growing interest in developing new assays and predictive tools. Science and technology are constantly evolving and the possibility of predicting the carcinogenicity of chemicals based on structure and/or molecular signatures is gaining attention. Although more accurate predictions of human safety will ultimately drive the application of these new tools, the pressure to reduce resources and minimize animal testing are also very real and directly contribute to the growing interest and application of alternative approaches. In Chap. 2, Lidya Stavitskaya, Jiri Aubrecht, and Naomi Kruhlak describe the current state of efforts by industry and FDA to use structure- and biology-based models to predict the mutagenicity and carcinogenicity of pharmaceuticals. The application of computational methods to evaluate the relationship between chemical structure and genotoxicity/carcinogenicity is relatively inexpensive and does not require actual chemical synthesis for testing. Therefore, quantitative structure-activity relationship (QSAR) models are being used more routinely in the early drug development process and for predicting mutagenicity of drug substance impurities. In contrast to QSAR, biology-based predictive models such as toxicogenomics are often used as investigative tools to address questions on the human relevance of findings. Whether these new biology-based models get incorporated into the standard mutagenicity/carcinogenicity testing paradigm either as replacements for any of the current studies or, more likely, as supplemental/supportive information will depend on further refinements, robust validation, and larger databases.

In Chap. 3, Laura Custer and Mark Powley describe the application and interpretation of the Ames assay as well as other in vitro and in vivo tests that are used to assess the potential genotoxicity of pharmaceuticals as described in International Conference on Harmonization (ICH) S2 (R1) [21]. [ICH is an organization involving regulators and research based industries from US, Europe, and Japan which was founded in 1990 to improve the efficiency of pharmaceutical R&D by developing and implementing harmonized guidelines and standards]. Some genotoxicity test results can be fairly straightforward. For example, a clear positive in the Ames assay would likely lead to a quick decision to terminate development of that compound. On the other hand, a small increase in micronuclei formation or chromosome aberrations relative to controls might require some follow up studies to put that finding into better perspective for human safety assessment. Laura Custer and Mark Powley review the different tests and strategies to de-risk these situations using a weight-of-evidence approach from both a regulatory and industry perspective.

In addition to the standard battery of genotoxicity tests that are required by regulatory authorities as outlined in ICH S2 (R1), there are number of new in vivo genotoxicity assays that are being developed to supplement and/or potentially substitute for the core battery. These new tests include the comet assay, the Pig-A gene mutation assay, and the liver micronucleus test [23–28]. In Chap. 4, Patricia Escobar, Stephen Dertinger, and Robert Heflich provide an overview of each of these tests, including their value and limitations as investigative tools in regulatory testing. The authors also briefly discuss strategies for de-risking positive findings in the core battery of genotoxicity assays and the re-emergence and interest in the transgenic rodent gene mutation assay for evaluating germ cell mutagenicity.

In addition to fully characterizing the genotoxicity and carcinogenicity of the active pharmaceutical ingredient, it is also necessary to assess the potential genotoxicity of process impurities and degradants. For context, all pharmaceutical impurities have to be identified, qualified, and controlled at certain threshold levels [29–31]. However, a lingering concern is that there may be genotoxic impurities below these threshold levels that could still lead to an increased and unacceptable risk for carcinogenicity. How to deal with these low level genotoxic impurities has been a challenging and frustrating issue for industry and regulatory scientists for a number of years largely due to different views on the overall safety risks and by the complexity of the technical and synthetic process changes that are often required to control them [32–34]. For example, it is well accepted that humans are exposed to naturally occurring carcinogens almost every single day of their lives through diet, lifestyle, and sunlight. So, what level of increased carcinogenic risk is considered negligible and how does that level of risk translate to a safe level of a mutagenic impurity? In addition, since the electrophilic nature (and inherent biological reactivity) of chemicals is highly variable and dependent on their unique structure, the potential carcinogenic risk cannot be the same for all mutagenic chemicals. So, recognizing that a global guidance was needed to standardize the criteria and control strategies for genotoxic impurities, industry and health authorities agreed to establish an ICH Expert Working Group (EWG) in 2010 to develop an international harmonized guideline. In Chap. 5, Peter Kasper and Lutz Muller, who were both members of this EWG discuss the history and concepts of the new and important ICH guidance on DNA reactive (mutagenic) impurities that was published in 2014 [35]. The authors also provide a few examples on how the principles of this guidance have been interpreted and applied in real world situations.

If the pharmaceutical industry can effectively screen out DNA reactive compounds and de-risk other potential genotoxic drugs with more sophisticated and relevant models, why is there a need to conduct 2-year rodent carcinogenicity studies and why are there so many positive findings in these studies, especially in the labels of approved drugs? Are tumor findings in rodents relevant to humans? While a deep dive into the mechanisms of non-genotoxic carcinogens is outside the scope of this book, it is clear that most drugs associated with tumors in animal studies are not DNA-reactive. For example, various hormones and growth factors can cause tumors in animals due to prolonged and exaggerated pharmacological effects at high doses [36–38]. Immunosuppressive drugs can lead to an increase in viral associated tumors in both animals and humans [39–41]. In fact, any drug that causes tissue hyperplasia in animals could be considered a suspect carcinogen (until proven otherwise) since increased cellular proliferation has long been recognized as a characteristic of tumor promotion and progression [42–44]. Therefore, it is not a question of whether there are non-genotoxic rodent carcinogens but rather are any of these considered relevant to humans. We know from decades of research in toxicology that many tumor findings in animals are not relevant to humans due to unique characteristics of rodent physiology and their subsequent response to chemicals. So, if we had readily accessible, sensitive, and specific biomarkers of carcinogenicity, humans could be monitored for these changes in clinical trials. Unfortunately, such biomarkers do not currently exist and the collection of most tissue samples from humans to investigate evidence of tissue hyperplasia is, of course, unreasonable. So, while there has been and continues to be an abundance of scientific debate on the predictive value of rodent carcinogenicity studies, especially in the pharmaceutical industry [45–50], rodent carcinogenicity studies are still conducted for most drugs, especially small molecules.

In Chap. 6, James MacDonald and David Jacobson-Kram provide a brief historical overview of the regulations regarding pharmaceutical safety testing, what we've learned from decades of rodent bioassay studies, what alternative testing approaches have been considered, and how carcinogenicity assessments may be refined in the future. The authors introduce an ongoing ICH initiative that is designed to test the ability of sponsors and Drug Regulatory Agencies (DRAs) to prospectively predict the outcome of 2-year carcinogenicity studies based on toxicology, pharmacology, and mechanistic endpoints. It is expected that a successful outcome of this exercise (i.e., the ability predict the results of carcinogenicity studies with a high degree of certainty) may lead to changes in the carcinogenicity testing requirements for small molecules in certain cases. More details on this ICH initiative are covered in Chap. 7. Other important topics covered in Chap. 6 include: (1) a description of how carcinogenicity study protocols and study results are reviewed by the FDA; (2) the role of the Carcinogenicity Assessment Committee (CAC) and Executive CAC in this process; and (3) the regulatory expectations in regards to the design and analysis of these studies.

In Chap. 7, Frank Sistare and Abby Jacobs cover four main topics including: (1) the current global regulatory requirements for carcinogenicity testing of small molecules and the limitations of these approaches; (2) numerous examples where positive rodent carcinogenicity study outcomes were not considered relevant to humans; (3) the increasing use of the 6-month transgenic rasH2 mouse model as part of the standard carcinogenicity testing paradigm; and (4) the ongoing effort within ICH to potentially reduce the number of 2-year rat carcinogenicity studies for small molecules. The global carcinogenicity testing requirements for small molecules is covered in a series of documents developed through ICH including ICH S1 (the need for long-term rodent carcinogenicity studies of pharmaceuticals) [51], ICH S1B (testing for carcinogenicity of pharmaceuticals) [9], and ICH S1C (R2) (dose selection for carcinogenicity studies) [52]. As mentioned previously, the ongoing initiative within ICH to potentially change the carcinogenicity testing paradigm for small molecules involves a prospective analysis of ongoing carcinogenicity studies by both sponsors and DRAs to determine how well the outcome of these studies can be predicted. The rationale for this initiative was supported by the results from a retrospective analysis of carcinogenicity studies conducted by the pharmaceutical industry which showed that almost 85 % of rat carcinogenicity study outcomes could be predicted by the mechanism of action of the drug and by the results from earlier nonclinical safety studies [53]. It was also estimated that almost 40 % of 2-year carcinogenicity studies could be avoided if no signals were detected after assessing these criteria. While this is a relatively high predictive value, especially considering the diverse range of drugs, it is not 100 % and a number of questions were raised by Health Authorities from US, Japan, and Europe. The main concern from regulators was about the 15 % of drugs that were not correctly predicted by this paradigm. So, the question to be answered was: if this process was followed, how many potential human carcinogens (false negatives) would "slip" through the system? Of course, the real answer is dependent upon whether one believes that any false negatives in the pharmaceutical industry data analysis represent true human carcinogens. In all these cases the drugs were approved anyway and, for most of the false negatives, there was a mechanistic explanation (e.g., species specific effect) or exposure margin that invoked no human relevance. Nevertheless, since regulators are charged with protecting human health, it is not hard to understand why any recommendation from the industry to eliminate the "gold standard" for carcinogenicity assessment would face some scrutiny.

However, despite the reluctance from health authorities to accept the industry proposal, the ability to predict the outcome of rat carcinogenicity studies is not without merit and the EMA, FDA, and PMDA ultimately agreed to participate in the prospective ICH study to test the industry hypothesis using a set of standardized criteria. It is expected this study will generate enough information in a real world situation so that health authorities can determine the ability of both sponsors and regulators to predict the outcome of the rat carcinogenicity study. Pending a successful outcome of this initiative, EMA, FDA, and PMDA agreed to consider revising ICH S1 and allow a waiver of 2-year rat carcinogenicity studies under certain circumstances.

In Chap. 8, Maggie Dempster et al. discuss the carcinogenicity testing of biopharmaceuticals which is included in ICH S6 (preclinical safety evaluation of biotechnology-derived pharmaceuticals) [54]. There are many distinct differences between small molecules and biopharmaceuticals with respect to their physicochemical and biological properties and these differences must be understood and appreciated in order to conduct the most appropriate carcinogenicity assessment. This is especially true for biopharmaceuticals that are not biologically active in rodents. In this chapter, Dempster et al. review the different classes of biopharmaceuticals such as growth factors and immunosuppressive drugs that have been associated with an increased tumorigenic risk in humans simply based on their pharmacology. The authors also present some case studies to show different approaches for evaluating the carcinogenic risk of biopharmaceuticals and in translating these findings to humans, including the use of pharmacovigilance data.

By necessity, 2-year carcinogenicity studies are conducted relatively late in drug development and neither sponsors nor regulators can afford to deal with inadequate studies or uninterpretable results just prior to registration. While the ICH S1 initiative may lead to a reduction in the overall number of 2-year carcinogenicity studies for small molecules, it will not eliminate them completely. In fact, it is estimated that for about half of all drugs, there will be enough uncertainty with respect to the predicted carcinogenicity outcome that a 2-year rat study will be required. So, the big question is what happens when you actually get a carcinogenic signal in your study? Does it matter? The answer depends on a number of factors including the strength of the tumor signal, the exposure margin (relative to the AUC at the recommended human dose), and the known relevance to humans. Regardless, a statistically significant tumor finding in a carcinogenicity study is a major event and can lead to unacceptable delays in development and marketing approval, and possibly even termination of the project.

In Chap. 9, Todd Bourcier and Denis Roy discuss what factors need to be considered in identifying and de-risking the human relevance of tumor findings in 2-year rat carcinogenicity studies. From the industry perspective, a positive signal in a 2-year rat carcinogenicity study has a huge business and financial impact, especially considering that these studies are generally conducted late in development to support marketing submissions. Given that the average development time of a drug is about 10 years and can cost > $2 billion dollars of R&D investments [55, 56], this is not the time to uncover major approvability issues. Of course, a positive rodent carcinogenicity study also puts regulators in a difficult situation since they do not want to withhold approvals of new medicines for a finding that may not be relevant to humans but, at the same time, they cannot take risks with protecting human health. Unfortunately, despite the best attempts to de-risk the carcinogenicity potential of new drugs, surprises do happen. At that point, the burden is mostly on industry scientists to propose a rationale scientific argument for why the finding is not likely relevant to humans. This may include conducting follow-up mechanistic and investigative studies to put the carcinogenicity finding into proper perspective and provide additional experimental evidence to support their hypothesis. Bourcier and Roy review the factors that need to be considered when designing and interpreting carcinogenicity studies. They also offer additional suggestions and guidance on how to manage and communicate carcinogenicity findings to internal and external stakeholders including a case study on the GLP-1 receptor agonists.

However, if the carcinogenicity studies are negative (i.e., no statistically significant increase in tumors in the treated groups), then the presumption would be that the molecule has essentially been de-risked as a carcinogen. (Note: some exceptions would include hormonal agents and immunosuppressive drugs where an increase in tumors may not be evident in the carcinogenicity studies, but where the concern for carcinogenicity may still exist based on the mechanism of action). Unfortunately, like all things in life, nothing is 100 % guaranteed and "stuff" happens. This is especially true in clinical trials where imbalances in tumor incidences can occur between treatment groups due to random chance and more rigorous medical examination of the subjects. The imbalance in tumor incidence can occur in either direction for the treatment group when compared to the controls but safety concerns are only raised when the incidence of a particular tumor in the treated group is increased. This is true even if there is no statistically significant difference in overall tumor incidence between the groups. As one can surmise, this is a very challenging situation and can lead to speculation of a possible tumor promotion effect. There are no well characterized or established tumor promotion models, and whether a "tumor promoter" can actually make it all the way through a toxicology program, including a clean carcinogenicity study without some signal is debatable, in and of itself. Nevertheless, this situation has happened more than once and in Chap. 10, Lorrene Buckley, Beatriz Silva-Lima, and Mark Tirmenstein discuss how a positive tumor signal in a clinical trial is determined and what kind of additional follow up investigative studies can be performed to further de-risk the concern. These follow up investigations must be designed on a case-by-case basis and with a very strong scientific rationale to fully interrogate biological plausibility, including an assessment of tumor promotion and progression. These latter assessments are especially critical given that tumors in clinical trials are not likely to arise de novo from drug treatment given the relatively short latency period. The authors briefly discuss a few models that can be used for studying tumor promotion and progression although it is well recognized that development of more relevant models is warranted. Finally, Buckley et al. present some important case studies in which clinical tumor findings were effectively de-risked by applying robust scientific arguments along with data from a few key follow-up investigative studies.

References

1. Gooderham NJ, Carmichael PL (2007) Mechanisms of chemical carcinogenesis. In: Alison MR (ed) The cancer handbook, 2nd edn. Wiley, Hoboken, New Jersey pp 1–16. doi:10.1002/0470025077
2. Loeb LA, Harris CC (2008) Advances in chemical carcinogenesis. Cancer Res 68:6863–6872. doi:10.1158/0008-5472.CAN-08-2852
3. Luch A (2005) Nature and nurture – lessons from chemical carcinogenesis. Nature Rev Cancer 5:113–125. doi:10.1038/nrc1546
4. Klaunig JE (2013) Chemical carcinogenesis. In: Klaassen CD (ed) Casarett and Doull's toxicology: the basic science of poisons, 8th edn. McGraw Hill, New York, pp 393–443
5. Tanaka T, Shimizu M, Kochi T, Moriwaki H (2013) Chemical-induced carcinogenesis. J Exp Clin Med 5:203–209. doi:10.1016/j.jecm.2013.10.009
6. Federal Food, Drugs, and Cosmetics Act, Drug Amendments of 1962. Public Law 87–781, 76 STAT 780 (1962)
7. Jacobs AC, Hatfield KP (2013) History of Chronic Toxicity and Animal Carcinogenicity Studies for Pharmaceuticals. Vet Pathol 50:324–333. doi:10.1177/0300985812450727
8. Weisburger JH, Weisburger EK (1967) Tests for animal carcinogens. In: Busch H (ed) Methods in cancer research. Academic, New York, pp 307–398
9. International Conference on Harmonization of Technical Requirements for Registration of Pharmaceuticals for Human Use (1997) ICH Harmonised Tripartite Guideline. Testing for carcinogenicity of pharmaceuticals S1B
10. Conney AH (1982) Induction of microsomal enzymes by foreign chemicals and carcinogenesis by polycyclic aromatic hydrocarbons: GHA Clowes Memorial Lecture. Cancer Res 42:4875–4917
11. Miller JA (1970) Carcinogenesis by chemicals: an overview – GHA Clowes Memorial Lecture. Cancer Res 30:559–576

12. Miller EC, Miller JA (1981) Searches for ultimate chemical carcinogens and their reactions with cellular macromolecules. Cancer 47:2327–2345. doi:10.1002/1097-0142(19810515)47:10<2327::AID-CNCR2820471003>3.0.CO;2-Z
13. Ames BN, McCann J, Yamasaki E (1975) Methods for detecting carcinogens and mutagens with the Salmonella/mammalian-microsome mutagenicity test. Mutat Res 31:347–364
14. Ashby J, Paton D (1993) The influence of chemical structure on the extent and sites of carcinogenesis for 522 rodent carcinogens and 55 different human carcinogen exposures. Mutat Res 286:3–74
15. Ashby J, Tennant RQ (1991) Definitive relationships among chemical structure, carcinogenicity and mutagenicity for 301 chemicals tested by the US NTP. Mutat Res 257:229–306
16. Gold LS, Slone TH, Stern BR, Bernstein L (1993) Comparison of target organs of carcinogenicity for mutagenic and non-mutagenic chemicals. Mutat Res 286:75–100
17. Kirkland D, Aardema M, Henderson L, Muller L (2005) Evaluation of the ability of a battery of three in vitro genotoxicity tests to discriminate rodent carcinogens and non-carcinogens. I. sensitivity, specificity, and relative predictivity. Mutat Res 584:1–256. doi:10.1016/j.mrgentox.2005.02.004
18. McCann J, Choi E, Yamasaki E, Ames BN (1975) Detection of carcinogens as mutagens in the Salmonella /microsome test; assay of 300 chemicals. Proc Natl Acad Sci U S A 72:5135–5139
19. Rosenkranz H, Ennever FK (1990) An association between mutagenicity and carcinogenicity potency. Mutat Res 244:61–65
20. Tennant RW, Margolin BH, Shelby MD, Zeiger E, Haseman JK, Spalding J, Caspary W, Resnick M, Stasiewicz S, Anderson B, Minor R (1987) Prediction of chemical carcinogenicity in rodents from in vitro genetic toxicity assays. Science 236:933–941
21. International Conference on Harmonization of Technical Requirements for Registration of Pharmaceuticals for Human Use (2011) ICH Harmonised Tripartite Guideline. Guidance on genotoxicity testing and data interpretation for pharmaceuticals intended for human use S2(R1)
22. MacGregor JT, Casciano D, Muller L (2000) Strategies and testing methods for identifying mutagenic risks. Mutat Res 455:3–20
23. Brendler-Schwaab S, Hartman A, Pfuhler S, Spelt G (2005) The in vivo comet assay; use and status in genotoxicity testing. Mutagenesis 20:245–254. doi:10.1019/mutage/gei033
24. Rothfuss A, O'Donovan M, De Boeck M, Brault D, Czich A, Custer L, Hamada S, Plappert-Helbig U, Hayashi M, Howe J, Kraynack AR, van der Leede B, Nakajima M, Priestley C, Thybaud V, Saigo K, Sawant S, Shi J, Storer R, Struwe M, Vock E, Galloway S (2010) Collaborative study on fifteen compounds in the rat-liver Comet assay integrated into 2- and 4-week repeat-dose studies. Mutation Res 702:40–69. doi:10.1016/j.mrgentox.2010.07.006
25. Bhalli J, Shaddock JG, Pearce MG, Dobrovolsky VN (2013) Sensitivity of the Pig-a assay for detecting gene mutation in rats exposed acutely to strong clastogens. Mutagenesis 28:447–455. doi:10.1093/mutage/get022
26. Dobrovolsky VN, Miura D, Heflich RH, Dertinger SD (2010) The in vivo Pig-a gene mutation assay, a potential tool for regulatory safety assessment. Environ Mol Mutagen 51:825–835. doi:10.1002/em.20627
27. Uno Y, Morita T, Luijten M, Beevers C, Hamada S, Itoh S, Ohyama W, Takasawa H (2015) Recommended protocols for the liver micronucleus test: report of the IWGT working group. Mutat Res 783:13–18
28. Sui H, Matsumoto H, Wako Y, Kawasako K (2015) Evaluation of in vivo genotoxicity by thioacetamide in a 28-day repeated-dose liver micronucleus assay using male young adult rats. Mutat Res 780–781:81–84
29. International Conference on Harmonization of Technical Requirements for Registration of Pharmaceuticals for Human Use (2006) ICH Harmonised Tripartite Guideline. Impurities in new drug substances Q3A(R2)

30. International Conference on Harmonization of Technical Requirements for Registration of Pharmaceuticals for Human Use (2006) ICH Harmonised Tripartite Guideline. Impurities in new drug products Q3B(R2)
31. International Conference on Harmonization of Technical Requirements for Registration of Pharmaceuticals for Human Use (2011) ICH Harmonised Tripartite Guideline. Impurities: guideline for residual solvents Q3C(R5)
32. Delaney EJ (2007) An impact assessment of the application of the threshold of toxicological concern concept to pharmaceuticals. Requl Toxicol Pharmacol 49:107–124. doi:10.1016/j.yrtph.2007.06.008
33. Jacobson-Kram D, Jacobs A (2005) Use of genotoxicity data to support clinical trials or positive genetox findings on a candidate pharmaceutical or impurity…now what? Int J Toxicol 24:129–134. doi:10.1080/10915810590952933
34. Snodin DJ, McCrossen SD (2013) Mutagenic impurities in pharmaceuticals: a critique of the derivation of the cancer TTC (threshold of toxicological concern) and recommendations for structural-class-based limits. Requl Toxicol Pharmacol 67:299–316. doi:10.1016/j.yrtph.2013.08.014
35. International Conference on Harmonization of Technical Requirements for Registration of Pharmaceuticals for Human Use (2014) ICH Harmonised Tripartite Guideline. Assessment and control of DNA reactive (mutagenic) impurities in pharmaceuticals to limit potential carcinogenic risk M7
36. Russo IH, Russo J (1998) Role of hormones in mammary cancer initiation and progression. J Mammary Gland Biol Neoplasia 3:49–61, 1083-3021/98/0100-0049
37. Ter Braak B, Siezen CL, Speksnijder E, Koedoot E, Van Steeg H, Salvatori DCF, Van de Water B, Van der Laan JW (2015) Mammary gland tumor promotion by chronic administration of IGF1 and the insulin analogue AspB10 in the p53R270H/+WAPCre mouse model. Breast Cancer Res 17:14
38. Vahle JL, Long GD, Sandusky G, Westmore M, Ma YL, Sato M (2004) Bone neoplasms in F344 rats given teriparatide [rhPTH(1–34)] are dependent on duration of treatment and dose. Toxicol Pathol 32:426–438. doi:10.1080/01926230490462138
39. Gutierrez-Dalmou A, Campistol JM (2007) Immunosuppressive therapy and malignancy in organ transplant recipients. Drugs 67:1167–1198, 0012-6667/07/0008-1167
40. Ponce RA, Gelzleichter T, Haggerty H, Heidel S, Holdren MS, Lebrec H, Mellon RD, Pallardy M (2014) Immunomodulation and lymphoma in humans. J Immunotoxicol 11:1–12. doi:10.3019/1547691X.2013.798388
41. Weaver JL (2012) Establishing the carcinogenic risk of immunomodulatory drugs. Toxicol Pathol 40:267–271. doi:10.1177/0192623311427711
42. Butterworth BE (1990) Consideration of both genotoxic and nongenotoxic mechanisms in predicting carcinogenic potential. Mutat Res 239:117–132, 0165-1110/90
43. Foster JR (2000) Cell death and cell proliferation in the control of normal and neoplastic tissue growth. Molecular Pathol 28:441–446, 0192-6233/00
44. Cohen SM, Arnold LL (2010) Chemical carcinogenesis. Toxicol Sci 120(S1):S76–S92. doi:10.1093/toxsci/kfq365
45. Alden CL, Lynn A, Bourdeau A, Morton D, Sistare FD, Kadambi VJ, Silverman L (2011) A critical review of the effectiveness of rodent pharmaceutical carcinogenesis testing in predicting human risk. Vet Pathol 48:772–784
46. Ames BN, Gold LS (1990) Chemical carcinogenesis; too many rodent carcinogens. Proc Natl Acad Sci U S A 87:7772–7776
47. Contrera J, Jacobs AC, DeGeorge JJ (1997) Carcinogenicity testing and the evaluation of regulatory requirements for pharmaceuticals. Requl Toxicol Pharmacol 25:13–145
48. Ennever FK, Lave LB (2003) Implications of the lack of accuracy of the lifetime rodent bioassay for predicting human carcinogenicity. Requl Toxicol Pharmacol 38:52–57. doi:10.1016/S0273-2300(03)00068-0
49. Jacobs A (2005) Prediction of 2-year carcinogenicity study results for pharmaceutical products: how are we doing? Toxicol Sci 88:18–23. doi:10.1093/toxsci/kfi248

50. Van Oosterhout JPJ, Van Der Laan JW, De Waal EJ, Olejniczak K, Hilgenfeld M, Schmidt V, Bass R (1997) The utility of two rodent species in carcinogenic risk assessment of pharmaceuticals in Europe. Requl Toxicol Pharmacol 25:6–17
51. International Conference on Harmonization of Technical Requirements for Registration of Pharmaceuticals for Human Use (1996) ICH Harmonised Tripartite Guideline. The need for long-term rodent carcinogenicity studies of pharmaceuticals S1A
52. International Conference on Harmonization of Technical Requirements for Registration of Pharmaceuticals for Human Use (2008) ICH Harmonised Tripartite Guideline. Dose selection for carcinogenicity studies S1C(R2)
53. Sistare FD, Morton D, Alden C, Christensen J, Keller D, De Jonghe S, Storer RD, Reddy MV, Kraynak A, Trela B, Bienvenu J-G, Bjurstrom S, Bosmans V, Brewster D, Colman K, Dominick M, Evans J, Hailey JR, Kinter L, Liu M, Mahrt C, Marien D, Myer J, Perry R, Potenta D, Roth A, Sherratt P, Singer T, Slim R, Soper K, Fransson-Steen R, Stoltz J, Turner O, Turnquist S, van Heerden M, Woicke J, DeGeorge JJ (2011) Toxicol Pathol 39:716–744. doi:10.1177/0192623311406935
54. International Conference on Harmonization of Technical Requirements for Registration of Pharmaceuticals for Human Use (2011) ICH Harmonised Tripartite Guideline. Preclinical safety evaluation of biotechnology-derived pharmaceuticals S6(R1)
55. Pharmaceutical Research and Manufacturers of America (2013) 2013 Biopharmaceutical Research Industry Profile. PhRMA, Washington, DC
56. Tufts Center for the Study of Drug Development (2014) Cost of developing a new drug. Tufts University, School of Medicine, Boston

Chapter 2
Chemical Structure-Based and Toxicogenomic Models

Lidiya Stavitskaya, Jiri Aubrecht, and Naomi L. Kruhlak

Abstract All pharmaceuticals undergo a comprehensive panel of non-clinical tests to ensure their safety before being approved for use in humans. Genetic toxicity and carcinogenicity testing in whole animal and cell-based assays are a significant component of this testing and the outcome of these studies can significantly impact the marketability of a drug product. Consequently, there is a strong desire for pharmaceutical developers to assess the potential of new drug candidates to cause these effects prior to investing significant resources and in advance of regulatory submission. Early screening of pharmaceutical candidates can be performed through the application of computational methods that evaluate the relationship between chemical structural descriptors and genotoxic and/or carcinogenic outcomes. (Quantitative) structure-activity relationship [(Q)SAR] models represent an entirely virtual assessment that relies only on knowledge of chemical structure to provide a prediction, making them inexpensive and rapid to apply without the need for synthesis of the potential drug candidate. Toxicogenomic-based approaches aim to assess biological functions based on interactions among various parts of biological systems with emphasis on molecular pathways. Toxicogenomics consists of interrogating pathways via studying the genome-wide gene expression changes, and is the most widely explored systems biology methodology in carcinogenicity and genotoxicity risk assessment. It is typically performed at later stages of the drug development process to interpret chemical exposure-induced perturbations of biological pathways by identifying molecular initiating events that can predict downstream genotoxic and carcinogenic consequences, as well as address the question of human

Disclaimer for FDA Authors The views expressed in this document have not been formally disseminated by the FDA and should not be construed to represent any agency determination or policy.

L. Stavitskaya • N.L. Kruhlak (✉)
US Food and Drug Administration, Center for Drug Evaluation and Research,
Silver Spring, MD 20993, USA
e-mail: Naomi.Kruhlak@fda.hhs.gov

J. Aubrecht
Drug Safety Research and Development, Pfizer Inc., Groton, CT 06355, USA

M.J. Graziano, D. Jacobson-Kram (eds.), *Genotoxicity and Carcinogenicity Testing of Pharmaceuticals*, DOI 10.1007/978-3-319-22084-0_2

relevance of non-clinical safety findings. Although the qualification of toxicogenomic-based approaches for regulatory use is being pursued by industry and regulatory agencies, the application of these methodologies is currently limited to exploratory analyses by pharmaceutical developers. In contrast, in addition to their application to early drug discovery, (Q)SAR models are now being routinely utilized under harmonized regulatory guidance for the late-stage safety assessment of drug substance impurities.

This chapter provides an overview of the use of chemical structure-based and toxicogenomic-based methods in pharmaceutical development, with a focus on the endpoints of genetic toxicity and carcinogenicity.

Keywords (Q)SAR • Modeling • Toxicogenomics • Risk-assessment • High-throughput

2.1 Introduction

All pharmaceuticals undergo a comprehensive panel of non-clinical tests to assess their safety. Historically, each of the three worldwide regulatory regions (USA, European Union, and Japan) designated their own battery of recommended tests to support market approval in their respective regions. However, due to the multinational nature of many pharmaceutical manufacturers, accommodating different regulatory expectations from different regions became cumbersome, driving the need for harmonized guidance. This is the premise behind the International Conference on Harmonisation (ICH) guidelines, which represents a set of tests and standards that is region-neutral and has been agreed to by all contributing regulatory organizations.

All drugs entering the modern-day regulatory review pipeline are required to undergo testing in a series of non-clinical studies, including *in vitro* and *in vivo* tests, as well as clinical trials to assess safety and efficacy. Among these tests are assays to determine the genotoxic and carcinogenic potential of the active pharmaceutical ingredient (API), as described by harmonized non-clinical safety guidance for these respective endpoints. Genetic toxicity testing is intended to identify the potential of a molecular entity to cause damage to DNA. Such a molecular initiating event is assumed to have implications for tumorigenicity and, consequently, a positive genotoxic outcome may be treated as an indicator of human carcinogenic potential. The full battery of genetic toxicity tests required for pharmaceuticals is described in ICH S2A and S2B, and is designed to assess mutagenicity, clastogenicity and direct DNA damage [1]. In general, positive genetic toxicity findings are unfavorable from a regulatory standpoint except when the risk-benefit ratio suggests otherwise, such as when the therapeutic candidate is intended for treatment of a life-threatening indication, treatment duration is very short and infrequent, and/or where the therapy is intended for an indication where no other treatment option exists. Although the standard genotoxicity testing battery has been shown to be highly sensitive for the detection of carcinogens, detecting 93 % in one estimate, the published literature also suggests that the testing paradigm features low specificity [2, 3]. For instance, 50 % of non-carcinogens among marketed pharmaceuticals had some positive genotoxicity findings in the ICH S2 standard test battery [4]. This reported discrepancy may partly be due to the limitations of selected genotoxicity assays, which show a high rate of false positives when predicting carcinogenicity, as well as the complexity of carcinogenic mechanisms (i.e., genotoxic vs. non-genotoxic). Furthermore, the observation that 50 % of chemicals tested in the 2-year rodent bioassay tested positive in at least one species or sex [5, 6] brings into question the relevance of tumors induced in rodents to humans [7–9]. Despite these limitations, *in vivo* testing to assess carcinogenic potential in rodents as a surrogate for human carcinogenesis has remained an important component of non-clinical safety assessment of new drugs.

The ICH S1A guideline [10], finalized in 1995, describes the specific conditions under which carcinogenicity testing of pharmaceuticals is recommended. In short, pharmaceuticals dosed for 6 months or longer require testing in the 2-year rodent bioassay. ICH S2A describes details of the conduct of the bioassay [1], and further information on the statistical interpretation of findings is documented in the Organization for Economic Cooperation and Development technical guidance, OECD TG-451 [11]. In contrast to the panel of required genetic toxicity tests, carcinogenicity testing is expensive, time consuming and uses large number of animals. The carcinogenicity study itself has a duration of 2 years with an additional year required for analysis of the harvested tissues from the approximately 500 study animals [12], representing a significant commitment from pharmaceutical manufacturers in both time and resources.

For both genetic toxicity and carcinogenicity testing, there is a strong desire by industry and regulatory agencies to be able to assess the potential of drug candidates to cause these effects early in the drug development process. The earlier such an assessment can be made, the lower the likelihood that a drug will be abandoned due to an undesirable toxicity profile and the lower the cost that will be incurred when taking such an action. The earliest screening activities undertaken by pharmaceutical developers use computer-based methodologies where a series of potential drug candidate structures are run against a battery of (quantitative) structure-activity relationship [(Q)SAR] models for toxicological or pharmacological endpoints, including genetic toxicity/rodent carcinogenicity, phospholipidosis, hERG channel inhibition, and hepatotoxicity through metabolism prediction [13], as well as calculated absorption, distribution, metabolism, and excretion properties [14]. The predictions from these models are then used to rank-order and prioritize chemical series for further optimization based on their desired pharmacological action and lack of potential toxicological liabilities. The advantages of *in silico*-based screening are many, including the lack of a need for synthesized test article, high-throughput, low cost, and a reduction in animal testing, as examples. The genotoxic potential of promising lead molecules is further evaluated using high throughput *in vitro* screening assays. Since the goal of screening is to predict the outcome of the standard genotoxicity testing battery, the screening assays typically mimic endpoints of the

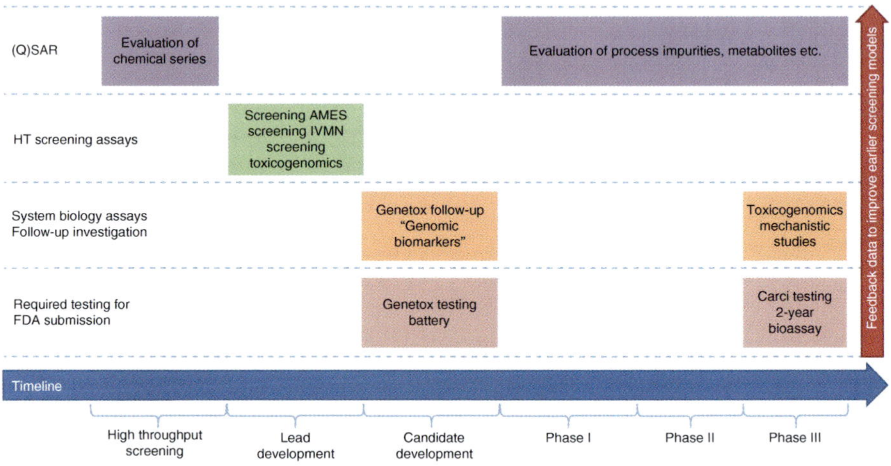

Fig. 2.1 Application of chemical computational- and genomic-based tools in drug development

genotoxicity testing battery [15]. These screens often include high-throughput versions of bacterial mutation assays to detect mutagenic compounds and modified *in vitro* micronucleus assays for detection of agents causing chromosome damage. In addition to their utility in confirming earlier assessments made using *in silico* [(Q)SAR] models, an advantage of these higher throughput *in vitro* assays is their potential to generate data that can in turn be used to enhance and refine the original (Q)SAR models used to assess new drug candidates entering the pipeline. Furthermore, (Q)SAR models are not only being used for safety assessment of an API, but are now being routinely used for the safety assessment of pharmaceutical impurities under the newly established ICH M7 guideline [16]. Under this guideline, (Q)SAR model predictions may be used to obviate the need for conventional *in vitro* testing under some circumstances, representing the first instance of their recommendation in harmonized regulatory guidance for this purpose. A timeline illustrating the application of chemical structure- and toxicogenomic-based tools in drug development is shown in Fig. 2.1.

Since the standard genotoxicity battery features high sensitivity and low specificity, follow-up safety testing strategies are needed to assess the risk and relevance of positive genotoxicity results to human health. This is of particular importance to *in vitro* chromosome damage assays, which tend towards lower specificity (higher false positive rates) for predicting carcinogenicity than other tests in the standard battery. Currently, the follow-up strategies consist of using similar DNA damage-sensing endpoints to provide further mechanistic information. These approaches typically include additional *in vitro* and *in vivo* assays for DNA damage such as the comet assay, histone H2AX phosphorylation assay, unscheduled DNA synthesis assay, and those addressing apoptosis and cell cycle perturbation (e.g., the TNEL assay or cell cycle analysis via FACS); however, these experimental strategies sometimes result in ambiguous outcomes and can raise more questions than they answer. Having a mechanistic understanding of positive genotoxicity findings would facilitate a more thorough assessment of human relevance with respect to overall carcinogenic risk and, as such, the development and acceptance of mechanistic-based tools is urgently needed. Recent progress in molecular technologies that enables the interrogation of responses to chemical exposure on the whole genome level (toxicogenomic approaches) provides new opportunities to study mechanisms of genotoxicity and carcinogenicity and has been evaluated by a variety of laboratories as a risk-assessment tool [17, 18]. If validated and adopted on a broad scale, toxicogenomics has the potential to significantly simplify follow-up testing and provide a catalyst for the development and acceptance of future screening assays.

2.2 Structure-Based Models

(Quantitative) structure-activity relationship [(Q)SAR] models describe the correlation between chemical structural features and biological activity under the general assumption that similar molecules exhibit similar physicochemical and biological properties [19]. The label "(Q)SAR" collectively refers to models that describe quantitative relationships (QSARs), where individual contributing features are weighted, and qualitative relationships (SARs), where the absence or presence of a feature is the basis of the prediction. The concept of QSARs was first reported by Hansch and Fujita in 1964 [20] describing the relationship between combinations of physical and chemical properties, and endpoints such as anesthetic potency and rodent carcinogenicity. Since that time, the approach has been expanded to a vast array of endpoints ranging from *in vitro* mutagenicity to clinical cardiovascular effects, utilizing thousands of different, though not necessarily independent, molecular descriptors. The early Corwin Hansch models (often termed "Hansch equations") were simple, mathematically-derived relationships that described the activity of a small number of highly similar structures. Nowadays, a range of complex mathematical algorithms are employed to derive such relationships using computational horsepower, and have the capability to identify correlations across large datasets containing thousands of diverse chemicals.

2.2.1 (Q)SAR Modeling Methodologies

Construction of a (Q)SAR model requires three fundamental components (Fig. 2.2): (1) a numerical representation of activity at the biological endpoint of interest for each member of a set of chemicals; (2) a chemical structure for each member of the set, represented by computer-interpretable molecular descriptors; and (3) an algorithm or method to extract relationships between the molecular descriptors and activity scores across the set [21].

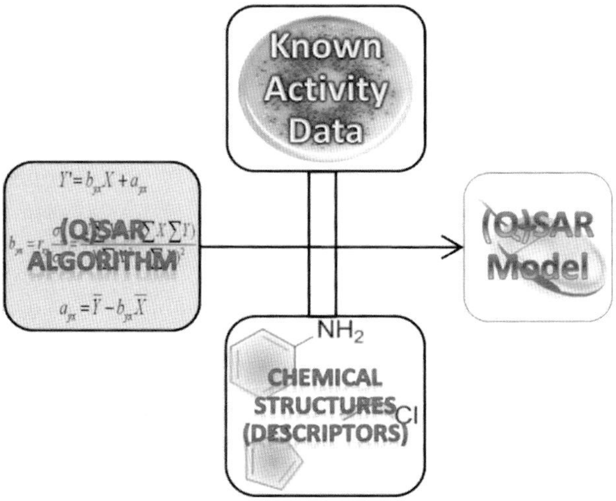

Fig. 2.2 Components required for the construction of a (Q)SAR model

Obtaining a good quality model depends heavily on the quality of data used to train the model, making data collection the first and most crucial step when constructing a model. For endpoints such as carcinogenicity and genotoxicity, when a clear threshold defining positive and negative responses exists, binary activity scores can be used to train the model, where a '0' denotes a non-toxicant and a '1' denotes a toxicant. A training set can be comprised of a few, highly structurally-similar chemicals to construct a local model or many, structurally-diverse chemicals to build a global model. Local models offer the advantage of providing high accuracy predictions for a specific class of chemicals. However, global models are often preferred due to their ability to make predictions for a variety of chemical classes, something that is particularly important for regulators encountering a broad range of chemicals attributes and structural motifs in submissions from pharmaceutical developers.

Molecular descriptors are representations of chemical structural features that can be interpreted by a computer. Model descriptors can be categorized as those that are explicit in the structure, specifically substructural fragments or functional groups, and those that are continuous numerical values, such as electrotopological descriptors calculated for each structure using mathematical equations [22]. While both types of descriptors are used widely in (Q)SAR modeling, interpretation of predictions generated by models based on complex calculated descriptors can be problematic as they are harder to relate to specific structurally alerting parts of molecules that can be readily interpreted from a chemistry perspective. In this regard, models constructed from more interpretable, fragment-based descriptors can be more desirable if the goal is to utilize the predictions to modify or to discover new lead structures in a series.

Chemical structure-based modeling methodologies can be characterized as being either statistical-based or expert rule-based, describing the way by which correlations between activity and substructural/whole-molecular attributes are identified. A broad range of statistical algorithms, such as linear regression, support vector machines, neural networks, random forests, and k-nearest neighbors can be used to develop a (Q)SAR model [23]. Statistical-based models use a machine-learning approach to identify a mathematical relationship between molecular features and biological activities, while an expert rule-based approach captures correlations derived by human experts to describe chemical motifs responsible for biological activity in the form of structural alerts, or nodes defining a decision tree. An example of a well-established structural alert that has been identified and encoded by both statistical and expert rule-based methodologies is the primary alkyl halide, which can cause bacterial mutagenesis by direct alkylation with DNA. While expert rule-based systems may use computers to assist in hypothesis generation, the alerts and subsequent knowledge base are encoded manually, providing a high degree of mechanistic detail and supporting citations. Statistical models offer the benefit of being rapid to construct and validate, even from extremely large and diverse datasets, but depending on the type of algorithm and descriptors that are used, they can vary widely in their interpretability rendering many of them black boxes. For some applications, such as regulatory use, this can make them difficult to apply. In contrast, expert rule-based systems can provide greater mechanistic interpretability in cases of positive predictions as a direct result of the extensive expert assessment used to develop them. The limitation of this, however, is that expert rule-based systems are relatively labor-intensive to develop because they require manual inspection of structurally similar analogs and extensive literature review to adequately characterize a structural alert [24, 25].

An additional consideration to the application of a model is that of applicability domain [26], which is the area of chemical space within which a model is able to make a reliable prediction. This area can be calculated based on the structural features and properties of the chemicals used to construct the model, whose values define the perimeter of chemical space and knowledge that the model has familiarity with. For a statistical-based model constructed from a classical (Q)SAR training set, a domain of applicability can be reasonably defined based on the structural attributes of the training set chemicals, for example, using molecular fragment representation or multi-dimensional chemical space; however, for an expert-rule based system, it can be harder to define such a domain since the model is based on collective knowledge rather than a finite set of training structures. This is particularly challenging in situations where a test chemical does not contain a structural alert or fall into an active category, since there may be little clarity between whether there is a lack of information about the chemical or whether definitive evidence exists that the chemical is negative. One approach to resolving this issue for expert systems is the use of a reference set of structures in combination with the prediction, where the known activity of compounds sharing structurally similar features can be reviewed and used for comparison purposes.

2.2.2 Model Validation

Predictive performance of a (Q)SAR model should ideally be assessed using a combination of statistical techniques such as non-cross-validation, cross-validation, y-scrambling, and external validation [23]. Non-cross-validation (or recall) is performed to examine the model's ability to predict chemicals within the training set using the final model. Whereas, during a cross-validation, a fraction of the chemicals is excluded from the training set, a model is rebuilt with the remaining chemicals, and then challenged with the excluded chemicals. The number of chemicals excluded in a cross-validation experiment can range from one (leave-one-out) to many (leave-many-out). Y-scrambling is an additional, internal validation method that provides baseline statistics from a model created from a randomized dataset to which other internal validation statistics can be compared. All three of these internal validation methods assess the robustness of a model. However, the predictivity of a model can only be directly assessed using an independent, external data set of "new" chemical structures with associated activity scores. If performed appropriately, external validation results can be indicative of real-world application. An external validation set should ideally be large, and representative of a broad range of chemical space, as well as be balanced in its ratio of empirically positive and negative chemicals. In general, a model's performance cannot be expected to exceed the reproducibility of the test data that it is constructed from.

The Organisation for Economic Co-operation and Development (OECD) has described a set of five criteria that it deems important for a (Q)SAR model to be considered suitable for regulatory purposes. These criteria are referred to as the OECD Validation Principles (previously known as the Setubal Principles) and include full statistical validation to assess robustness and predictivity, as described above, as well as other broader recommendations addressing model relevance and transparency. The ICH M7 guideline outlines the use of (Q)SAR models for assessing the genotoxic potential of drug impurities and references the OECD Validation Principles specifically, stating that (Q)SAR models used under the guideline "… should follow the general validation principles set forth by the Organisation for Economic Co-operation and Development (OECD)" [16].

The OECD Validation Principles are as follows:

To facilitate the consideration of a (Q)SAR model for regulatory purposes, it should be associated with the following information:

1) *a defined endpoint*
2) *an unambiguous algorithm*
3) *a defined domain of applicability*
4) *appropriate measures of goodness-of–fit, robustness and predictivity*
5) *a mechanistic interpretation, if possible*

Principle 1: The endpoint for which the model provides a prediction should be well-defined and should be relevant to the regulatory endpoint to which the model is being applied (e.g., bacterial mutagenicity and 2-year rodent carcinogenicity).

Furthermore, the data the model were trained upon will directly represent the endpoint for which the model can make a prediction.

Principle 2: The use of transparent algorithms (e.g., linear regression, support vector machine, neural networks, random forest, and k-nearest neighbors) is emphasized to enable independent assessment of a model's decision-making process by regulators and the broader scientific community. Freely available (Q)SAR models are more often associated with detailed documentation supporting their construction and application. In contrast, commercially available (Q)SAR models may lack complete algorithmic transparency due to their position in a competitive marketplace; however many commercial vendors make an effort to publish their methodologies in the peer-reviewed scientific literature to provide credibility to their approaches.

Principle 3: The domain of applicability represents the area(s) of biological and/or chemical space within which a model has existing knowledge, and within which it can reliably be used to make a prediction. The area is defined by the attributes (descriptors) used to construct the model, and the relative values of those attributes for a given test compound determine whether that compound falls within the applicability domain. Examples of methods used to assess whether a compound is within the applicability domain include calculating the number and proximity of its nearest neighbors in chemical space, or determining whether its chemical fragments are present within the training set structures.

Principle 4: As described in detail above, a model should be fully characterized using internal validation (e.g., non-cross- and cross-validation) and external validation techniques to provide evidence of its ability to represent the data upon which it was trained, as well as its ability to provide predictions for new compounds. A good model is one that uses a minimal, but optimal, number of molecular descriptors to describe the training dataset without over-fitting the data.

Principle 5: Where possible, it is desirable for mechanistic associations within a model to be documented, allowing interpretation of predictions from a biological and/or chemical standpoint. For example, the bacterial mutagenic potential of an aromatic nitro group can be rationalized through its ability to form a nitrenium ion capable of reacting with DNA. This principle can be fulfilled by determining if there is scientific evidence of known mechanistic basis to the (Q)SAR and using descriptors that were previously determined to have mechanistic rationale [27].

2.2.3 (Q)SAR Models for Regulatory Non-clinical Safety

Regulatory research in the development of chemical structure-linked data repositories and (Q)SAR models for pharmaceutical non-clinical safety endpoints has been underway for many years [28–34] While this technology continues to be used routinely in other regulatory areas [35], such as for the safety assessment of environmental chemicals [36, 37] or food-contact substances [38], the exposure levels encountered for pharmaceuticals combined with the need for high predictive accuracy have necessitated a cautious adoption of these methodologies for product review and approval purposes, where protecting patient safety remains paramount. For pharmaceuticals, early regulatory applications of (Q)SAR models and cheminformatics techniques were on an investigational basis only, for hypothesis generation or subsequently for regulatory decision support, where models contributed to the weight-of-evidence supporting a regulatory action. Carcinogenicity and genetic toxicity were identified as endpoints for which there was both a regulatory interest as well as an adequate amount of data available from standardized protocols for model building and knowledge development. An emphasis was placed on modeling methods that provided prediction transparency and some degree of chemical or biological interpretability due to the need to interrogate the rationale behind predictions. Over time, genetic toxicity emerged as an area where the mechanisms of activity based on chemical reactivity with DNA became reasonably well characterized and sufficient amounts of high-quality data became available for both model development and validation. Carcinogenicity model development has lagged behind due to the greater complexity of biological mechanisms involved, as well as the overall lower availability of data; however, it remains an area of active research and is potentially the next endpoint for regulatory consideration when applying (Q)SAR models to pharmaceuticals.

2.2.3.1 Genotoxicity

Harmonized regulatory guidance now endorses the use of (Q)SAR models for the genetic toxicity assessment of pharmaceutical impurities. Under the recently finalized ICH M7 guideline, bacterial mutagenicity assessment of impurities may be performed through the use of two (Q)SAR methodologies, one statistical-based and one expert rule-based, that predict the outcome of a bacterial mutation (Ames) assay [16]. The guideline states that if neither methodology generates a positive prediction, then the impurity is considered qualified and no further testing is required. Conversely, if either one of the methodologies generates a positive prediction, then the impurity is treated as being potentially mutagenic unless further evidence suggests otherwise. Indeed, the guideline contains a provision for the use of expert knowledge, which may be used to support or overturn a prediction. The premise behind the combined use of two (Q)SAR methodologies is that they demonstrate complementarity through the use of different structural descriptors and algorithms, thereby providing a more robust assessment of mutagenic activity in the absence of empirical data. This approach is supported by several external validation studies that demonstrate the benefit of using combinations of models to generate an overall prediction [39–41] where the parameters of negative predictivity and sensitivity can be raised to acceptably high levels (~80 and ~90 %, respectively, in one example). This reflects a reduction in false negative predictions, supporting the regulatory imperative to protect patient safety, particularly for a drug impurity that offers no therapeutic benefit. As expected, this also leads to an increase in false positive predictions; however, this effect can be mitigated to some degree through the use of expert knowledge, which, from a practical standpoint, is typically easier to apply to positive predictions rather than negative ones. Overall, interpreting model output with the use of expert knowledge has been shown to provide additional improvement to performance statistics in external validation studies [40, 41]. This process can include assessing a structurally alerting functionality to determine its biological plausibility, determining that the prediction (either positive or negative) is based upon an adequate number of structurally relevant analogs in the training set, or considering empirical data for structurally related analogs that were not used to train the model, as examples.

One of the key considerations when applying expert knowledge to a model prediction is the need to be able to interpret a prediction in terms of possible reaction chemistry and supporting training set structures. Furthermore, the ability to identify a "structural alert" from a model prediction is underscored by the provision in the ICH M7 guideline for overturning a positive prediction for an impurity if the prediction is based on an alert that is shared with the API, and the API has been shown to be empirically negative. In general, interpretability, as well as predictive performance of a model may affect its suitability for application under ICH M7.

To remain vendor neutral, as well as to allow for the use of novel, in-house (Q)SAR methodologies that may be used by pharmaceutical developers, the ICH M7 guideline does not specify the use of any particular software, but simply recommends that the models meet the general definition of statistical or rule-based methodologies. There are many commercial and open source bacterial mutagenicity models that fall into these general categories, but substantially less that offer the interpretability that enables detailed evaluation of model predictions facilitating the application of expert knowledge. For models that regulatory agencies have limited experience with, additional information about their construction, validation, and training sets may be required to support their application under ICH M7 to demonstrate quality and reliability of their predictions.

Although the content of a (Q)SAR report may differ depending on the complexity of the prediction, all reports should ideally include key information such as the version of software and models used, raw model predictions, and an explanation of any conclusions that are based on expert interpretation of the (Q)SAR data, particularly if those conclusions differ from the raw model output. A well-performed and well-documented (Q)SAR assessment under ICH M7 reduces the likelihood that additional review-related questions will arise.

2.2.3.2 Carcinogenicity

Although not as widely accepted by regulators, the use of (Q)SAR models for carcinogenicity may increase as a replacement for conventional toxicology testing of pharmaceuticals due to its relatively low cost in time and resources, its consistency in performing assessments, and its transparency of predictions. The scope of the 2-year rodent bioassay has evolved since its introduction in 1995, with provisions for testing in transgenic mice being introduced in 1998 [42]. More recently, the bioassay has been challenged by the so-called "NegCarc" initiative, which proposes a battery of shorter-duration assays including 6–12-month toxicology studies and genetic toxicity tests. The absence of a positive finding from this alternative battery is sufficient to conclude that the compound lacks carcinogenic potential, with the exception of compounds causing hormonal perturbation, which are excluded from the proposal altogether. It has been estimated that this strategy could help eliminate 40 % of 2-year rat studies from the drug development process [43].

The main barrier to regulatory acceptance of (Q)SAR for predicting the carcinogenic potential of pharmaceuticals has undoubtedly been its lower predictive accuracy in the past, particularly for non-genotoxic carcinogens. Recently, FDA/CDER demonstrated that predictive performance of 64–78 % in sensitivity and 72–85 % in negative predictivity could be obtained when applying global (Q)SAR models in combination when applied to external validation sets comprised of drug-like and industrial chemicals [43, 44]. While these new models still do not demonstrate the same predictive performance as seen with those for genetic toxicity, they offer a significant improvement over earlier published results and support the possibility that they may in the future contribute to the weight-of-evidence for regulatory decision-making on the carcinogenic potential of new drugs.

2.2.4 Screening vs. Regulatory Use of (Q)SAR

The use of (Q)SAR models to support regulatory decision-making for pharmaceuticals has been discussed in detail in Sect. 2.2.3. In addition to this application, (Q)SAR models are commonly used by pharmaceutical developers during early discovery and development phases to eliminate candidates with genotoxic and, by extension, potential carcinogenic liabilities. In many cases, the same models may be utilized for these different applications but tuned to have optimal predictive characteristics for each purpose (Table. 2.1). For example, the sensitivity/specificity trade-off may be adjusted in favor of high sensitivity for late-stage regulatory use, where false negatives are undesirable as they compromise patient safety, in contrast to high specificity for early candidate screening, where false positives are undesirable as they result in the unnecessary elimination of promising pharmaceutical candidates.

Table 2.1 Summary of the relative attributes of models used for drug candidate screening and regulatory support

Property	Drug candidate screening	Regulatory support
Timeframe	Early	Late
Predictivity	Low false positive rate	Low false negative rate
Expert analysis	Sometimes	Yes
Throughput	High	Low
Prediction transparency	Not necessarily	Yes
Molecular descriptors	Any	Explainable

Similarly, a battery of models may be used to increase sensitivity by applying an "any positive equals a positive overall" rule, or a "majority wins" rule may be used to increase specificity. In some instances, however, particular models may not be suitable for both applications. As discussed earlier, it is desirable that (Q)SAR models used for regulatory purposes provide transparency and interpretability in their predictions to facilitate the use of expert knowledge to provide additional supportive evidence for overall conclusions. This may translate to the need for model descriptors that allow a prediction to be explained in terms of chemical reactivity or biological mechanisms, as well as result in a lower-throughput process overall. In contrast, for models used for early screening purposes, the onus is on high-throughput analyses, where the lack of need for expert interpretation of predictions provides greater flexibility in the types of model algorithms and descriptors that can be used. In such a situation, models with the best predictive characteristics may be favored regardless of whether they essentially function as black boxes. Taken a step further, many pharmaceutical developers train their screening models on in-house proprietary data since it may be the most relevant to the area of chemical space in which they are developing new candidates. In contrast, the need for transparency in models used for regulatory support generally outweighs the benefits of proprietary training sets and, in general, models tend to be more heavily based on non-proprietary data. Furthermore, this allows the same models to be available to both regulators and developers, providing not only transparency in predictions but transparency in subsequent regulatory decision-making based upon these predictions.

2.3 Toxicogenomics

Toxicogenomic approaches develop knowledge on biological functions based on studying interactions among various parts of biological systems, via genome-wide monitoring of gene expression. The emphasis on pathways and their relationships to biological processes differentiates the field of toxicogenomics from traditional approaches that are mostly based on understanding the function of a single gene and/or protein. Because molecular pathways – not necessarily individual genes – are generally conserved across species from yeast to human [45], system biology approaches have the potential to bridge *in vitro* and *in vivo* preclinical studies with clinical investigations. Furthermore, differences in pathways among species can provide valuable insights in understanding molecular mechanisms of species-specific responses to chemical exposures.

The field of toxicogenomics has been enabled by a wide application of microarray technologies capable of interrogating gene expression (mRNA levels) on the whole genome level. Fundamental assumptions of toxicogenomics are that all toxicological effects are accompanied by gene expression changes [46], and that similar toxicological mechanisms cause comparable expression changes. Several successful applications of toxicogenomics to risk assessment have included the identification of hazards by comparing gene expression profiles of unknown compounds with a database of profiles derived from toxicants of known mechanisms of action [47, 48], the generation of mechanistic information via pathway analysis revealing biological processes affected by exposures [49–52], and subsequent prediction of a limited number of specific adverse effects [53–56].

2.3.1 Toxicogenomic Methodologies

Microarrays are available in a variety of designs, yet those based on oligonucleotides, either printed on a solid support, attached to beads, or synthesized *in situ* onto a wafer chip, have gained more popularity then printed cDNA arrays, due to standardization of production and high reproducibility. Currently, high quality microarrays with reproducible performance and high sensitivity are readily available for toxicogenomic studies including standardized analytical protocols [57]. Furthermore, recent progress in Next Generation Sequencing (NGS) technologies provides several advantages over microarrays, including a highly sensitive open system that enables the identification of novel transcripts, splice variants, microRNA, and other noncoding regulatory elements in one sequencing run. The standardization of NGS methods and procedures has been addressed by an international project called Sequencing Quality Control (SEQC) modeled after the successful Microarray Quality Control (MAQC) [58].

In general, a toxicogenomic study consists of three major components: (1) The biological model, (2) the assay platform, and (3) data analysis and interpretation. Toxicogenomic approaches are generally applied as mechanistic or predictive tools. Mechanistic toxicogenomic approaches consist of analyzing gene expression profiles for perturbation of molecular pathways and their relation to possible toxic mechanisms. Predictive toxicogenomics relies on creation of a gene expression profiles database following treatment of a biological model with reference agents (i.e., compounds with known mechanism of toxicity) and generation of marker gene sets, also called gene signatures or classifiers, that are associated with particular toxic effects. This approach aims to predict or classify unknown agents with respect to their potential toxicity or mode of action based on the presence or absence of a particular gene signature [47, 59]. Recently, the application of gene expression signatures, also called genomic biomarkers, in genetic safety risk assessment has been published by Li et al. [60] and is described in more detail in Sect. 2.3.2.1 using caffeine as a case study.

2.3.2 Toxicogenomics in Genotoxicity and Carcinogenicity Testing

Genotoxicity testing has been widely used as a surrogate for carcinogenicity assessment because of the mechanistic connection of DNA damage to cancer, and the practical limitations of the rodent 2-year carcinogenicity bioassay. However, it is well documented that genetic toxicity assays provide a specific but not sensitive measure of carcinogenic potential due to the variety of other mechanistic pathways by which tumorigenesis could occur, including those that are immune- or hormone-mediated. Therefore identification of key events and a mode of action (MOA) in rodent bioassays may provide a more rational basis for human cancer hazard and risk assessment [61] as a supplement to information gleaned from genomic biomarkers describing genotoxic mechanisms. This provides an opportunity for system biology approaches, specifically the mechanistic and predictive application of toxicogenomics.

2.3.2.1 Genotoxicity

The genotoxicity testing battery provides essential information on the potential of compounds to cause DNA damage. Since a single mutation in a specific oncogene has been shown to cause cancer, it is assumed that any exposure to DNA-reactive entities such as alkylating agents poses a cancer risk. On the other hand, it is commonly accepted that some genotoxic agents exhibit a thresholded dose–response curve [62] implying that exposures below the threshold are cancer risk free. Examples of mechanisms producing thresholded responses include DNA non-reactive interactions such as inhibition of enzymes involved in protein and DNA synthesis or DNA repair, inhibition of Na/K transport, inhibition of topoisomerases, and inhibition of processes leading to an imbalance of DNA precursors, energy depletion, production of active oxygen species, lipid peroxidation, and nuclease release from lysosomes [63]. Thus differentiating between DNA-reactive and DNA non-reactive mechanisms of genotoxicity by understanding their underlying mechanisms is critical [64]. Unfortunately, assessment of the risk and relevance of positive findings, particularly in the *in vitro* chromosome damage assays, poses a major challenge to industry and regulatory agencies. Follow-up strategies often have uncertain outcomes and rely on additional studies that use DNA damage-sensing endpoints. This has been recognized as an opportunity for toxicogenomic approaches [50, 65].

The basic concept for differentiation of genotoxic mechanisms via toxicogenomics involves (a) developing a database of gene expression profiles representing well-characterized molecular pathways using agents that elicit a wide range of cellular stresses; (b) developing a gene signature (genomic biomarker) that can differentiate DNA-reactive from non-reactive stresses; and (c) evaluating whether the genomic biomarker is present in a gene expression profile of cells treated with the tested chemical. The data from several laboratories and international consortia efforts (HESI and CarcinoGenomics) showed the ability of genomic biomarkers to differentiate DNA reactive and DNA non-reactive mechanisms of genotoxicity in TK6 [49, 66, 67], L5178Y [51, 66], and HepG2 and HepaRG cells [68–70].

The application of genomic biomarkers in genetic safety risk assessment has been proposed as a follow-up to positive findings in the *in vitro* chromosome damage assay (Fig. 2.3) [71]. Toxicogenomic biomarkers would provide mechanistic insights by

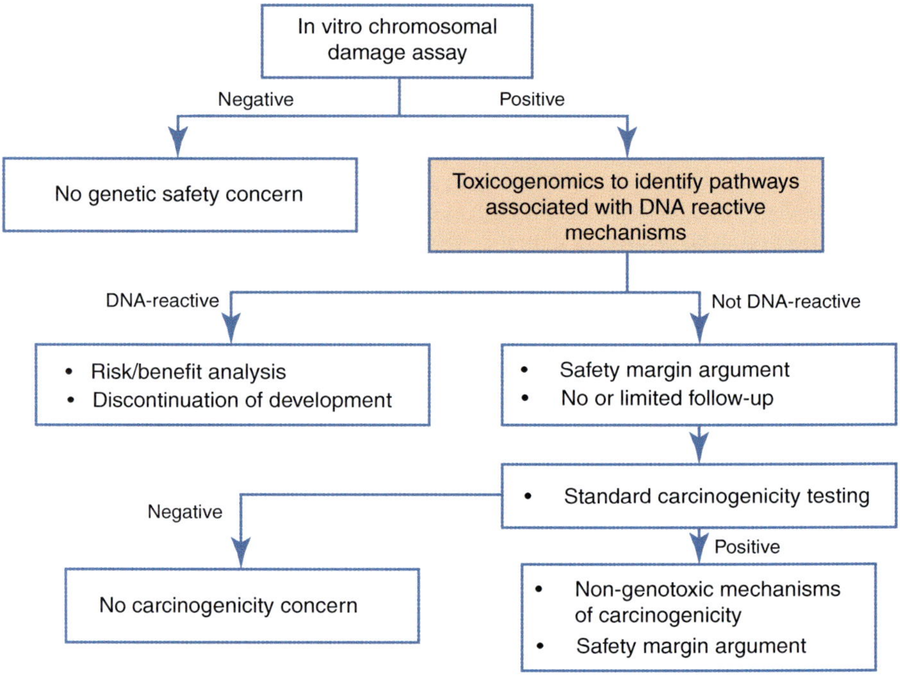

Fig. 2.3 Proposed incorporation of genomic biomarkers into genetic toxicology risk assessment (Reproduced with minor modifications from Ref. [60] with permission)

differentiating DNA-reactive from non-DNA-reactive mechanisms. For example, chemicals directly interacting with DNA by forming DNA adducts trigger the genomic signature associated with a genotoxic stress response and no safe level of exposure would be assumed. In contrast, non-DNA-reactive mechanisms devoid of a genotoxicity signature feature a threshold dose–response with no-effect levels that can be used for risk management. The utility of the genomic biomarker for de-risking or identifying irrelevant positives in chromosome damage assays was demonstrated via a Voluntary Genomics Data Submission (VXDS) that was discussed with the US Food and Drug Administration (FDA) using caffeine as a case study [60, 71]. In this study, caffeine served as a prototypical example of compound with well documented positive findings in the in vitro chromosome damage assays with no documented carcinogenicity risk to humans. The application of genomic biomarkers clearly demonstrated the absence of DNA damage pathway activation suggesting that the *in vitro* positive chromosome damage findings are of non-DNA-reactive mechanisms and are irrelevant to cancer risk to humans [60]. It was proposed that application of the genomic biomarker approach could eliminate the need for currently used *in vitro* and *in vivo* follow-up assays. The feedback from the FDA [65] was the major impetus to advance the formal qualification of the genomic biomarker via HESI under the Genomic Biomarker Qualification Project [60], where the project scope was expanded to include a large set of agents with known mechanisms of genotoxicity. Qualification of such a biomarker approach is anticipated to limit the need for additional *in vivo* testing and thereby simplify genetic safety risk assessment for drugs and chemicals and significantly reduce the need for animal testing.

2.3.2.2 Carcinogenicity

The rodent carcinogenicity bioassay has been used for over 30 years to assess the human cancer risk of chemicals. It requires a lifetime exposure of rats and mice (~18–24 months) with a test compound up to a maximum tolerated dose. Because of the extensive resources required, only a small fraction of chemicals have undergone carcinogenicity testing. In addition to resource constraints and ethical concerns, the high doses often used in the bioassay and the species differences between rodents and humans have led to considerable debate over the relevance of rodent carcinogenicity findings and provided impetus for the development of alternative methods, including attempts to exploit toxicogenomic-based methods. In contrast to the expectation that genotoxic chemicals are carcinogens, non-genotoxic chemicals cannot be assumed to be non-carcinogens. Thus, there has been a strong emphasis on understanding the chemical's mode of action to better translate the risk and relevance of rodent carcinogenicity findings to humans [72, 73]. The basic premise of toxicogenomic analysis in carcinogenicity testing is that gene expression changes in the target tissue precede and/or contribute to tumor development and these changes can be monitored after a short-term *in vivo* treatment to predict longer-term carcinogenic outcomes. This hypothesis has been extensively evaluated and numerous genomic biomarkers or signatures have been described to predict rat hepatocarcinogenicity induced by genotoxic and non-genotoxic compounds [53–56, 74, 75]. Recently, methods enabling a quantitative dose–response analysis from genomics data have been developed [76, 77]. This concept utilizes a transcriptional benchmark dose (BMD) that is applicable beyond a single target organ. This approach might be useful for non-selective chemicals that perturb many pathways, since this BMD value estimates doses at which the system begins to be perturbed by the toxicant [77]. In case of selective chemicals with a narrowly defined mechanism of action designed to perturb only a limited number of pathways (i.e. drug candidates), toxicogenomics might be used to provide detailed insights into molecular mechanisms that will aid in risk assessment by potentially deriving BMDs for molecular initiating events. This approach has recently been published as a case study using benzo(a)pyrene as a model agent [78].

Although these approaches are promising, additional detailed studies are needed in order to fully characterize the potential of toxicogenomics in carcinogenicity testing.

2.4 Future Perspectives

(Q)SAR models are widely used in the pharmaceutical industry for structure-based lead discovery and optimization. More recently, with the release of the ICH M7, the use of genetic toxicity (Q)SAR models during late-stage safety assessment is becoming more prevalent, where models for bacterial mutagenicity are considered fit-for-purpose for use in place of empirical *in vitro* testing for a pharmaceutical impurity. While (Q)SAR models may not yet be sufficiently mature to use prospectively as the basis of regulatory safety decisions for APIs, they can make a valuable contribution to the overall weight-of-evidence supporting a regulatory decision. As such, the endpoint of carcinogenicity is a logical one for the future regulatory application of (Q)SAR models given the recent interest in replacing the current testing paradigm, which requires the two-year rodent bioassay. Overall, (Q)SAR tools are very rapid to apply and increasingly reliable especially when they provide a high degree of interpretability and transparency, making them excellent safety assessment tools for both regulators and pharmaceutical developers alike.

Toxicogenomics provides a new avenue for interrogating mechanisms of genotoxicity and carcinogenicity. Currently, the evaluation of a variety of genomic biomarkers is being pursued by industry and regulatory agencies. Although more data are needed, toxicogenomic biomarkers of genotoxic stress responses have the potential to provide mechanistic context to positive findings in the *in vitro* chromosome damage assays. Furthermore, the transcriptional BMD may find future application in the direct characterization of cancer risk associated with chemical exposure.

The development and expansion of robust, high quality data sets for training (Q)SAR models will undoubtedly remain a priority and will likely serve to support the acceptance of (Q)SAR models for regulatory use. Similar challenges are encountered by regulators and pharmaceutical developers with respect to data quality and accessibility. Both parties possess vast amounts of toxicological information that could be used for modeling, but often data are not archived in a format amenable to automated knowledge extraction. Discussions are currently underway to facilitate sharing of selected proprietary findings from pharmaceutical developers to fill known data gaps in both (Q)SAR models and toxicogenomic data sets. This will improve overall model performance, as well as the increase the number of new chemicals for which a model is able to make a reliable prediction. As new and more unique drug candidates are designed, the ability to maintain up-to-date and relevant training sets and knowledge remains critical, as does the need to continue to push the state-of-the-science forward to develop increasingly sophisticated and predictive methodologies to be applied to the next generation of therapeutics.

References

1. ICH (2011) S2(R1) Guidance on genotoxicity testing and data interpretation for pharmaceuticals for human use. http://www.ich.org/fileadmin/Public_Web_Site/ICH_Products/Guidelines/Safety/S2_R1/Step4/S2R1_Step4.pdf
2. Kirkland D, Aardema M, Muller L, Hayashi M (2006) Evaluation of the ability of a battery of three in vitro genotoxicity tests to discriminate rodent carcinogens and non-carcinogens II. Further analysis of mammalian cell results, relative predictivity and tumour profiles. Mutat Res 608(1):29–42
3. Kirkland D, Aardema M, Henderson L, Muller L (2005) Evaluation of the ability of a battery of three in vitro genotoxicity tests to discriminate rodent carcinogens and non-carcinogens I. Sensitivity, specificity and relative predictivity. Mutat Res 584(1–2):1–256

4. Snyder RD, Green JW (2001) A review of the genotoxicity of marketed pharmaceuticals. Mutat Res 488(2):151–169
5. Hoffmann S, Hartung T (2006) Toward an evidence-based toxicology. Hum Exp Toxicol 25(9):497–513
6. MacDonald JS (2004) Human carcinogenic risk evaluation, part IV: assessment of human risk of cancer from chemical exposure using a global weight-of-evidence approach. Toxicol Sci 82(1):3–8
7. Knight A, Bailey J, Balcombe J (2006) Animal carcinogenicity studies: 2. Obstacles to extrapolation of data to humans. Altern Lab Anim 34(1):29–38
8. Vinken M, Doktorova T, Ellinger-Ziegelbauer H, Ahr HJ, Lock E, Carmichael P et al (2008) The carcinoGENOMICS project: critical selection of model compounds for the development of omics-based in vitro carcinogenicity screening assays. Mutat Res 659(3): 202–210
9. Thomas RS, O'Connell TM, Pluta L, Wolfinger RD, Yang L, Page TJ (2007) A comparison of transcriptomic and metabonomic technologies for identifying biomarkers predictive of two-year rodent cancer bioassays. Toxicol Sci 96(1):40–46
10. ICH (1995) S1A Guideline on the need for carcinogenicity studies of pharmaceuticals. http://www.ich.org/fileadmin/Public_Web_Site/ICH_Products/Guidelines/Safety/S1A/Step4/S1A_Guideline.pdf
11. OECD (2008) Test Guideline 451: carcinogenicity studies. http://www.oecd.org/chemical-safety/testing/41753121.pdf
12. Jacobson-Kram D, Sistare FD, Jacobs AC (2004) Use of transgenic mice in carcinogenicity hazard assessment. Toxicol Pathol 32(Suppl 1):49–52
13. Muster W, Breidenbach A, Fischer H, Kirchner S, Muller L, Pahler A (2008) Computational toxicology in drug development. Drug Discov Today 13(7–8):303–310
14. Leeson PD, Springthorpe B (2007) The influence of drug-like concepts on decision-making in medicinal chemistry. Nat Rev Drug Discov 6(11):881–890
15. Aubrecht J, Xu JJ (2010) Screening approaches for genetic toxicology. In: Xu JJ, Urban L (eds) Predictive toxicology in drug safety. Cambridge University Press, New York, pp 18–33
16. ICH (2014) M7 Assessment and control of DNA reactive (mutagenic) impurities in pharmaceuticals to limit potential carcinogenic risk. http://www.ich.org/fileadmin/Public_Web_Site/ICH_Products/Guidelines/Multidisciplinary/M7/M7_Step_4.pdf
17. Ku WW, Aubrecht J, Mauthe RJ, Schiestl RH, Fornace AJ Jr (2007) Genetic toxicity assessment: employing the best science for human safety evaluation Part VII: Why not start with a single test: a transformational alternative to genotoxicity hazard and risk assessment. Toxicol Sci 99(1):20–25
18. Waters MD, Jackson M, Lea I (2010) Characterizing and predicting carcinogenicity and mode of action using conventional and toxicogenomics methods. Mutat Res 705(3):184–200
19. Hendrickson JB (1991) Concepts and applications of molecular similarity – Johnson, Ma, Maggiora, Gm. Science 252(5009):1189
20. Hansch C, Fujita T (1964) Additions and corrections – analysis. A method for the correlation of biological activity and chemical structure. J Am Chem Soc. 86(24):5710
21. Tropsha A (2010) Best practices for QSAR model development, validation, and exploitation. Mol Inform 29(6–7):476–488
22. Todeschini R, Consonni V (2009) Molecular descriptors for chemoinformatics (2 vols). Wiley-VCH, Weinheim, 1257 p
23. OECD (2007) Guidance document on the validation of (Quantitative) structure activity relationships [(Q)SAR] models. http://www.oecd.org/officialdocuments/publicdisplaydocument pdf/?doclanguage=en&cote=env/jm/mono%282007%292
24. Ridings JE, Barratt MD, Cary R, Earnshaw CG, Eggington CE, Ellis MK et al (1996) Computer prediction of possible toxic action from chemical structure: an update on the DEREK system. Toxicology 106(1–3):267–279
25. Benigni R, Bossa C (2008) Structure alerts for carcinogenicity, and the salmonella assay system: a novel insight through the chemical relational databases technology. Mutat Res 659(3):248–261

26. Eriksson L, Jaworska J, Worth AP, Cronin MT, McDowell RM, Gramatica P (2003) Methods for reliability and uncertainty assessment and for applicability evaluations of classification- and regression-based QSARs. Environ Health Perspect 111(10):1361–1375
27. Cronin MT, Madden JC (2010) In silico toxicology: principles and applications. Chapter 11: Characterisation, evaluation and possible validation of in silico models for toxicity: Determining if a prediction is valid. Royal Society of Chemistry 275–300
28. Contrera JF, Matthews EJ, Kruhlak NL, Benz RD (2005) In silico screening of chemicals for bacterial mutagenicity using electrotopological E-state indices and MDL QSAR software. Requl Toxicol Pharmacol 43(3):313–323
29. Contrera JF, Matthews EJ, Kruhlak NL, Benz RD (2008) In silico screening of chemicals for genetic toxicity using MDL-QSAR, nonparametric discriminant analysis, E-state, connectivity, and molecular property descriptors. Toxicol Mech Methods 18(2–3):207–216
30. Kruhlak NL, Benz RD, Zhou H, Colatsky TJ (2012) (Q)SAR modeling and safety assessment in regulatory review. Clin Pharmacol Ther 91(3):529–534
31. Kruhlak NL, Contrera JF, Benz RD, Matthews EJ (2007) Progress in QSAR toxicity screening of pharmaceutical impurities and other FDA regulated products. Adv Drug Deliv Rev 59(1):43–55
32. Matthews EJ, Contrera JF (1998) A new highly specific method for predicting the carcinogenic potential of pharmaceuticals in rodents using enhanced MCASE QSAR-ES software. Requl Toxicol Pharmacol 28(3):242–264
33. Matthews EJ, Kruhlak NL, Cimino MC, Benz RD, Contrera JF (2006) An analysis of genetic toxicity, reproductive and developmental toxicity, and carcinogenicity data: I. Identification of carcinogens using surrogate endpoints. Requl Toxicol Pharmacol 44(2):83–96
34. Matthews EJ, Kruhlak NL, Cimino MC, Benz RD, Contrera JF (2006) An analysis of genetic toxicity, reproductive and developmental toxicity, and carcinogenicity data: II. Identification of genotoxicants, reprotoxicants, and carcinogens using in silico methods. Requl Toxicol Pharmacol 44(2):97–110
35. Bailey AB, Chanderbhan R, Collazo-Braier N, Cheeseman MA, Twaroski ML (2005) The use of structure-activity relationship analysis in the food contact notification program. Requl Toxicol Pharmacol 42(2):225–235
36. Cronin MT, Walker JD, Jaworska JS, Comber MH, Watts CD, Worth AP (2003) Use of QSARs in international decision-making frameworks to predict ecologic effects and environmental fate of chemical substances. Environ Health Perspect 111(10):1376–1390
37. Walker JD (2003) Applications of QSARs in toxicology: a US government perspective. J Mol Struct (THEOCHEM) 622(1–2):167–184
38. Arvidson KB, Chanderbhan R, Muldoon-Jacobs K, Mayer J, Ogungbesan A (2010) Regulatory use of computational toxicology tools and databases at the united states food and drug Administration's office of food additive safety. Expert Opin Drug Metab Toxicol 6(7):793–796
39. Hillebrecht A, Muster W, Brigo A, Kansy M, Weiser T, Singer T (2011) Comparative evaluation of in silico systems for ames test mutagenicity prediction: scope and limitations. Chem Res Toxicol 24(6):843–854
40. Sutter A, Amberg A, Boyer S, Brigo A, Contrera JF, Custer LL et al (2013) Use of in silico systems and expert knowledge for structure-based assessment of potentially mutagenic impurities. Requl Toxicol Pharmacol 67(1):39–52
41. Contrera JF (2013) Validation of Toxtree and SciQSAR in silico predictive software using a publicly available benchmark mutagenicity database and their applicability for the qualification of impurities in pharmaceuticals. Requl Toxicol Pharmacol 67(2):285–293
42. ICH (1998) S1B Testing for carcinogenicity of pharmaceuticals. http://www.ich.org/fileadmin/Public_Web_Site/ICH_Products/Guidelines/Safety/S1B/Step4/S1B_Guideline.pdf
43. Bourcier TM, McGovern TJ, Stavitskaya L, Kruhlak NL, Jacobson Kram D (2015) Improving prediction of carcinogenicity while aiming to reduce, refine, and replace the use of experimental animals. J Am Assoc Lab Anim Sci 54(2)):163–169
44. Stavitskaya L, Kruhlak NL, Cross KP, Minnier BL, Bower DA, Chakravarti S et al (2013) Development of improved in silico models for predicting rodent carcinogenicity. Int J Toxicol 32(1):72

45. Ideker T, Galitski T, Hood L (2001) A new approach to decoding life: systems biology. Annu Rev Genomics Hum Genet 2:343–372
46. Farr S, Dunn RT 2nd (1999) Concise review: gene expression applied to toxicology. Toxicol Sci 50(1):1–9
47. Hamadeh HK, Bushel PR, Jayadev S, DiSorbo O, Bennett L, Li L et al (2002) Prediction of compound signature using high density gene expression profiling. Toxicol Sci 67(2):232–240
48. Briede JJ, van Delft JM, de Kok TM, van Herwijnen MH, Maas LM, Gottschalk RW et al (2010) Global gene expression analysis reveals differences in cellular responses to hydroxyl- and superoxide anion radical-induced oxidative stress in caco-2 cells. Toxicol Sci 114(2):193–203
49. Amundson SA, Do KT, Vinikoor L, Koch-Paiz CA, Bittner ML, Trent JM et al (2005) Stress-specific signatures: expression profiling of p53 wild-type and -null human cells. Oncogene 24(28):4572–4579
50. Aubrecht J, Caba E (2005) Gene expression profile analysis: an emerging approach to investigate mechanisms of genotoxicity. Pharmacogenomics 6(4):419–428
51. Dickinson DA, Warnes GR, Quievryn G, Messer J, Zhitkovich A, Rubitski E et al (2004) Differentiation of DNA reactive and non-reactive genotoxic mechanisms using gene expression profile analysis. Mutat Res 549(1–2):29–41
52. Heinloth AN, Shackelford RE, Innes CL, Bennett L, Li L, Amin RP et al (2003) Identification of distinct and common gene expression changes after oxidative stress and gamma and ultraviolet radiation. Mol Carcinog 37(2):65–82
53. Ellinger-Ziegelbauer H, Aubrecht J, Kleinjans JC, Ahr HJ (2008) Application of toxicogenomics to study mechanisms of genotoxicity and carcinogenicity. Toxicol Lett 186(1):36–44
54. Ellinger-Ziegelbauer H, Stuart B, Wahle B, Bomann W, Ahr HJ (2005) Comparison of the expression profiles induced by genotoxic and nongenotoxic carcinogens in rat liver. Mutat Res 575(1–2):61–84
55. Fielden MR, Brennan R, Gollub J (2007) A gene expression biomarker provides early prediction and mechanistic assessment of hepatic tumor induction by nongenotoxic chemicals. Toxicol Sci 99(1):90–100
56. Fielden MR, Adai A, Dunn RT 2nd, Olaharski A, Searfoss G, Sina J et al (2011) Development and evaluation of a genomic signature for the prediction and mechanistic assessment of nongenotoxic hepatocarcinogens in the rat. Toxicol Sci 124(1):54–74
57. Shi L, Campbell G, Jones WD, Campagne F, Wen Z, Walker SJ et al (2010) The MicroArray Quality Control (MAQC)-II study of common practices for the development and validation of microarray-based predictive models. Nat Biotechnol 28(8):827–838
58. Consortium SM-I (2014) A comprehensive assessment of RNA-seq accuracy, reproducibility and information content by the sequencing quality control consortium. Nat Biotechnol 32(9):903–914
59. Hamadeh HK, Bushel PR, Jayadev S, Martin K, DiSorbo O, Sieber S et al (2002) Gene expression analysis reveals chemical-specific profiles. Toxicol Sci 67(2):219–231
60. Heng-Hong L, Hyduke DR, Chen R, Heard P, Aubrecht J, Fornace AJ Jr (2015) Development of a toxicogenomics signature for genotoxicity using a dose-optimization and informatics strategy in human cells. Environ Mol Mutagen 56(6):505–519
61. Holsapple MP, Pitot HC, Cohen SM, Boobis AR, Klaunig JE, Pastoor T et al (2006) Mode of action in relevance of rodent liver tumors to human cancer risk. Toxicol Sci 89(1):51–56
62. Henderson L, Albertini S, Aardema M (2000) Thresholds in genotoxicity responses. Mutat Res 464(1):123–128
63. Scott D, Galloway SM, Marshall RR, Ishidate M Jr, Brusick D, Ashby J et al (1991) International commission for protection against environmental mutagens and carcinogens. Genotoxicity under extreme culture conditions. A report from ICPEMC task group 9. Mutat Res 257(2):147–205
64. Kirkland DJ, Muller L (2000) Interpretation of the biological relevance of genotoxicity test results: the importance of thresholds. Mutat Res 464(1):137–147
65. Thybaud V, Le Fevre AC, Boitier E (2007) Application of toxicogenomics to genetic toxicology risk assessment. Environ Mol Mutagen 48(5):369–379

66. Akerman GS, Rosenzweig BA, Domon OE, McGarrity LJ, Blankenship LR, Tsai CA et al (2004) Gene expression profiles and genetic damage in benzo(a)pyrene diol epoxide-exposed TK6 cells. Mutat Res 549(1–2):43–64
67. Islaih M, Li B, Kadura IA, Reid-Hubbard JL, Deahl JT, Altizer JL et al (2004) Comparison of gene expression changes induced in mouse and human cells treated with direct-acting mutagens. Environ Mol Mutagen 44(5):401–419
68. Doktorova TY, Yildirimman R, Vinken M, Vilardell M, Vanhaecke T, Gmuender H et al (2013) Transcriptomic responses generated by hepatocarcinogens in a battery of liver-based in vitro models. Carcinogenesis 34(6):1393–1402
69. Jennen DG, Gaj S, Giesbertz PJ, van Delft JH, Evelo CT, Kleinjans JC (2010) Biotransformation pathway maps in WikiPathways enable direct visualization of drug metabolism related expression changes. Drug Discov Today 15(19–20):851–858
70. Jetten MJ, Kleinjans JC, Claessen SM, Chesne C, van Delft JH (2013) Baseline and genotoxic compound induced gene expression profiles in HepG2 and HepaRG compared to primary human hepatocytes. Toxicol In Vitro 27(7):2031–2040
71. Goodsaid FM, Smur S, Aubrecht J, Burczynski ME, Catalano J, Carl K (2010) Voluntary exploratory data submissions to the USFDA and the EMA: experience and impact. Nat Rev Drug Discov 9:435–445
72. Jacobs A (2005) Prediction of 2-year carcinogenicity study results for pharmaceutical products: how are we doing? Toxicol Sci 88(1):18–23
73. Jacobs A, Jacobson-Kram D (2004) Human carcinogenic risk evaluation, Part III: Assessing cancer hazard and risk in human drug development. Toxicol Sci 81(2):260–262
74. Ellinger-Ziegelbauer H, Stuart B, Wahle B, Bomann W, Ahr HJ (2004) Characteristic expression profiles induced by genotoxic carcinogens in rat liver. Toxicol Sci 77(1):19–34
75. Fielden M, Nelson B, Kherani A (2011) Acute intraocular inflammation following intravitreal injection of bevacizumab – a large cluster of cases. Acta Ophthalmol 89(8):e664–e665
76. Thomas RS, Allen BC, Nong A, Yang L, Bermudez E, Clewell HJ 3rd et al (2007) A method to integrate benchmark dose estimates with genomic data to assess the functional effects of chemical exposure. Toxicol Sci 98(1):240–248
77. Adler S, Basketter D, Creton S, Pelkonen O, van Benthem J, Zuang V et al (2011) Alternative (non-animal) methods for cosmetics testing: current status and future prospects-2010. Arch Toxicol 85(5):367–485
78. Moffat I, Chepelev N (2015) Comparison of toxicogenomics and traditional approaches to inform mode of action and points of departure in human risk assessment of benzo[a]pyrene in drinking water. Crit Rev Toxicol 45:1–43

Chapter 3
Genotoxicity Testing of API

L.L. Custer and M.W. Powley

Abstract Fortunately the frequency of mutagenic drug candidates recognized in early drug development is very low because positive Ames bacterial mutagenicity test results usually have severe ramifications with APIs dropped from further development. While negative results in the appropriate follow-up tests (e.g., mouse carcinogenicity studies) could enable progression of API development, pragmatically the cost and long duration required for these tests means that most pharmaceutical companies drop mutagenic APIs from development and quickly move on to another drug candidate.

Positive in vitro mammalian genotoxicity tests are far more frequently encountered but most times do not prevent further development of an API. When positive in vitro mammalian genotoxicity test results are obtained, follow-up studies in complementary in vitro or in vivo assays are quickly conducted to understand the biological relevance of the initial positive response. In the majority of cases the additional testing allows the sponsor to demonstrate lack of biological relevance or establish a safety threshold (generally ≥10×) based on exposures in animals at the no observed genotoxic effect level (NOGEL) relative to highest clinical exposure. For all these reasons, follow-up testing in response to an in vitro mammalian cell positive test is generally pursued and often results in continued development of the API.

This chapter focuses on genotoxicity testing strategies and includes case studies where follow-up studies were used to effectively de-risk positive in vitro genotoxicity test results. The regulatory guidelines (e.g., ICH S2(R1)) dictating which tests are required and which follow-up tests are acceptable are also presented.

Disclaimer The views expressed are those of the author. No official support or endorsement by the US Food and Drug Administration is provided.

L.L. Custer (✉)
Drug Safety Evaluation, Bristol-Myers Squibb, New Brunswick, NJ, USA
e-mail: laura.custer@bms.com

M.W. Powley
U.S. Food and Drug Administration, Center for Drug Evaluation and Research,
Silver Spring, MD, USA

M.J. Graziano, D. Jacobson-Kram (eds.), *Genotoxicity and Carcinogenicity Testing of Pharmaceuticals*, DOI 10.1007/978-3-319-22084-0_3

Because the regulatory guidelines list acceptable follow-up tests with little to no guidance on how to select them, this chapter provides recommendations on how to design the most appropriate follow-up testing strategy.

Keywords Genotoxicity • Ames bacterial mutagenicity test • DNA damage • Clastogenicity • ICH S2(R1) • No-observed genotoxic effect level (NOGEL)

3.1 Introduction

Genotoxicity is the process by which chemicals interact with or damage DNA and/or the cellular apparatus which regulates the accuracy and efficiency of the DNA replication and repair processes [UK COM 2000]. Many in vitro and in vivo tests have come and gone over the years but they all measure mutation, chromosome damage, DNA damage or repair, or numeric change in chromosome number. Genotoxicity testing has been used since the 1970s as a quick surrogate for long-term rodent carcinogenicity studies. This testing was initiated due to concerns regarding a chemical's direct potential to induce somatic cell mutations responsible for cancer initiation or indirect modes of action responsible for cancer progression. Initially, germ-cell genotoxicity assays were considered important for assessing potential chemical mutations that could have far reaching implications for future generations. Today, only somatic cell genotoxicity tests are required since they identify all known germ-cell genotoxicants.

In the last decade there has been a dramatic move from small molecule drugs created via organic synthesis to large molecule biologics (antibodies, peptides, proteins) harvested from cultured cells and combinations such as antibody drug conjugates. Genotoxicity testing is required for small molecules but not for the majority of biologicals. For this reason, testing strategies and recommendations in this chapter will refer to small molecules.

Once a drug candidate is selected for possible clinical development, it is subjected to an extensive battery of genotoxicity tests conducted under good laboratory practices (GLP) using protocols accepted by international regulatory authorities. These studies are submitted to regulatory health authorities as part of the overall nonclinical safety data package to support most Phase 1 clinical trials. In the US, the nonclinical safety data to support Phase 1 clinical trials are summarized in an Investigational New Drug (IND) application. Since new drugs are often tested first in healthy human volunteers who receive no benefit from the drug, the investigational drug must pose a very low health risk to those volunteers. Drugs are not evaluated for potential carcinogenicity until much later in drug development, so genotoxicity testing is conducted prior to Phase 1 clinical trials to protect human volunteers from exposure to potential carcinogens. Genotoxicity tests evaluate endpoints associated with initiation or progression of tumorigenesis using short-term in vitro and in vivo assays. These tests measure mutations, DNA strand breaks, chromosome damage, and changes in chromosome numbers.

3.2 Mechanisms of Genotoxicity

Genotoxic effects can arise through multiple mechanisms. From a broad perspective these effects can be categorized as direct or indirect genotoxicity. Direct genotoxicity is due to the interaction of a chemical API and/or its metabolite(s) with DNA. In contrast, indirect effects are associated with a chemical or metabolite(s) interacting with non-DNA targets and subsequent generation of genetic damage. An understanding of mechanism is critical for assessing clinical risks as well as determining appropriate strategies for following-up testing to further characterize positive results.

3.2.1 Direct Genotoxicity

The direct genotoxic mechanisms of primary concern in drug development are the induction of gene mutations and structural chromosomal damage (i.e., clastogenicity). Gene mutations include base-pair substitutions and frameshift mutations arising through addition or deletion of a single nucleotide. Clastogens cause DNA strand breaks which, if not repaired properly, can lead to the addition, deletion, or rearrangement of genetic information. Because mechanisms of direct genotoxicity have been associated with the multi-step process of carcinogenicity, positive results in assays capable of detecting gene mutations or clastogens are a significant concern for clinical trial subjects.

3.2.2 Indirect Genotoxicity

There are many examples of genotoxicity that result from indirect mechanisms of action. Of particular importance to drug development is an alteration in the number of an individual chromosomes. This process, known as aneugenicity, is the product of mitotic spindle apparatus disruption leading to the loss or gain of a whole chromosome. Another numerical chromosome effect is polyploidy, duplications of the entire chromosomal complement. Aneugenicity is associated with carcinogenicity and is a critical component of genotoxicity testing to support drug development. Additional indirect genotoxic mechanisms include inhibition of DNA synthesis, nucleotide pool imbalance, inhibition of topoisomerase, and in vivo body temperature changes.

3.2.3 Thresholds

Thresholds are defined as the dose or exposure above which an effect becomes discernable from background. Thresholds can be an important consideration when genetic toxicology testing yields positive results. Although there is mounting scientific evidence to suggest otherwise, genotoxicity associated with a direct mechanism is often treated as though no threshold exists. This outcome is likely related to (1) the complex nature of establishing thresholds for test articles capable of inducing gene mutations or clastogenic effects and/or (2) the negative outlook for developing genotoxic drugs.

In contrast, thresholds have been identified for indirect acting genotoxic compounds. When indirect mechanisms are suspected, it may be possible to identify no observed genotoxic effect levels (NOGELs) in appropriate in vivo testing. Development of a drug may proceed when acceptable margins of safety are shown to exist.

3.3 Regulatory Recommendations

The primary regulatory guideline addressing the genetic toxicology testing of small molecule drugs is the International Conference for Harmonization (ICH) S2(R1) guideline (ICH, 2011). This globally adopted guideline provides recommendations regarding the battery of genetic toxicology tests needed to support drug development (see below). In addition, the document describes the proper conduct of assays (e.g., test conditions, appropriate test systems, doses, etc.), evaluation of results, and appropriate follow-up for positive findings.

3.3.1 ICH Guidelines

Although quantitative risk assessment may be possible, genetic toxicology assays used to support drug development are primarily intended as hazard identification tools. Because there is no single genetic toxicology assay capable of effectively evaluating all genotoxic mechanisms, ICH S2(R1) recommends conducting a battery of tests to maximize the potential of identifying relevant effects. Due to the established association with carcinogenicity, the effects of most concern in drug development are gene mutations, structural chromosomal damage (i.e., clastogenicity), and numerical chromosomal damage (e.g., aneugenicity). The guideline offers flexibility by providing two testing options.

The Option 1 battery includes in vitro assays intended to evaluate gene mutations and chromosomal damage as well as an in vivo evaluation of chromosome level effects. The potential to induce gene mutations is evaluated through the bacterial reverse-mutation assay (i.e., Ames assay). In vitro chromosomal damage can be evaluated using the chromosomal aberration assay, micronucleus assay, or mouse lymphoma assay. All three of these assays are considered equivalent for regulatory use. In addition to the overall positive or negative assay result, additional information may be obtained to establish mechanism of action. For instance, the in vitro micronucleus assay can distinguish between clastogens and aneugens through the use of special staining procedures or flow cytometry. Likewise, mechanistic information can be gained from the mouse lymphoma assay through sizing of mutant colonies. Option 1 also requires an in vivo assay. This is most often the micronucleus assay in rodent hematopoietic cells; however, the chromosomal aberration assay is also acceptable. As with the in vitro assay, mechanistic data can be used to support the identification of clastogenic vs. aneugenic drugs.

Option 2 also involves conduct of an in vitro bacterial reverse mutation assay but requires in vivo evaluation of two genotoxic endpoints in two tissues. A possible combination to satisfy the in vivo requirements would be the micronucleus assay in rodent hematopoietic cells and the Comet assay. While the liver may be the default target organ for the Comet assay, understanding of metabolism and target organ toxicity can be used to justify selection of other tissue(s) as well.

Details of the regulatory genetic toxicology assays, including sources of standardized protocols (e.g., OECD guidelines), are summarized in Table 3.1 below. Following standardized protocols assures experiments are conducted according to widely accepted methodology, a critical consideration for regulatory use.

ICH M3(R2) describes timelines for submitting genetic toxicology data to support small molecule development. Under Option 1 the in vitro assays should be submitted prior to a Phase 1 clinical trial and the in vivo assay prior to Phase 2. While not explicitly stated in ICH M3(R2), a general assumption is that all assays will be submitted prior to initiating Phase 1 clinical trials when Option 2 is selected. A more limited evaluation, in some cases requiring no genetic toxicology data, may be acceptable for exploratory clinical trials. The timing of assays conducted to follow-up a positive result will vary. However, it is worth noting that, in some instances, a positive genotoxicity finding may result in a clinical hold until the risk is mitigated. Note that the timing of genetic toxicology testing is different for anticancer drugs. Per ICH S9, results of genotoxicity testing need only be performed to support submission of a marketing application.

ICH S6 (R1) addresses the safety evaluation of biological drugs such as proteins, peptides, monoclonal antibodies, etc. Standard genotoxicity testing is usually unnecessary as these molecules are unlikely to react with DNA. An exception is when an organic moiety is used to link 2 moieties together (e.g., therapeutic protein and PEG).

3.3.2 FDA Guidance

In addition to the ICH S2(R1) guideline, other sources of regulatory recommendations regarding genetic toxicology testing are available. For instance, the U.S. Food and Drug Administration (FDA) provides guidance on the evaluation and integration of genetic toxicology data (FDA, 2006). The FDA also addresses genetic

Table 3.1 Regulatory genetic toxicology assays

Assay	Damage detected	Standard methods[a]
ICH S2(R1)		
Option 1 and 2 *in vitro* assays		
Bacterial reverse mutation assay	Gene mutations	OECD 471
Chromosomal aberration assay	Structural and numerical chromosomal damage	OECD 473
Mouse lymphoma assay	Gene mutations Structural and numerical chromosomal damage Note: mutant colony sizing can be used to differentiate between mechanisms Gene mutations (large colonies) Clastogenic effects (small colonies)	OECD 476
Micronucleus assay	Structural and numerical damage Note: application of special staining procedures or use of flow cytometry can be used to differentiate between clastogenic and aneugenic effects	OECD 487
ICH S2(R1)		
Option 1 and 2 *in vivo* assays		
Comet assay	Primary DNA damage (e.g., single and double strand breaks)	OECD 489
Chromosomal aberration assay	Structural and numerical chromosomal damage	OECD 475
Micronucleus assay	Described above	OECD 474
Transgenic mouse gene mutation assay	Gene mutations (all tissues)	OECD 488
Other *in vivo* assays		
DNA covalent binding assay	Primary DNA damage (i.e., DNA adducts)	N/A
Liver Unscheduled DNA Synthesis (UDS) assay	Primary DNA damage (e.g., DNA repair)	OECD 486
Pig-a gene mutation assay	Gene mutations (red blood cells)	N/A

[a]OECD protocols can be found at: http://www.oecd-ilibrary.org/environment/oecd-guidelines-for-the-testing-of-chemicals-section-4-health-effects_20745788

toxicology evaluation of drug metabolites formed uniquely or disproportionately in humans (FDA, 2008). When these metabolites are encountered, the initial evaluation should include in vitro data for point mutations and chromosomal aberrations. No further testing is needed if in vitro results are negative; however, additional data may be warranted to further characterize genotoxic potential if positive results are obtained.

3.4 Pre-candidate Genotoxicity Screening

With few exceptions, mutagenicity is considered an unacceptable risk and companies usually drop the drug candidate from development due to the extensive amount of work required to demonstrate safety prior to initiating human clinical trials and the low probability of success. Mechanistic data may be used to demonstrate that mutagenicity is not relevant to human safety (e.g. bacterial specific metabolite, rat S9 specific metabolite, etc.). But aside from oncology drugs, and drugs targeting serious or life threatening diseases, most companies will not invest in these de-risking strategies and will quickly move on to another drug candidate. Therefore, pharmaceutical companies have implemented screening strategies during discovery and lead compound optimization to quickly identify mutagens as well as other serious safety liabilities. Screening strategies vary greatly between companies with no consensus on the best test(s) to use or the exact timing on when these tests should be conducted. However, the majority of pharmaceutical companies use a common strategy of first assessing the chemical structure for the presence of known mutagenic and carcinogenic structural alerts. If a decision is not made to eliminate the compound based on the structure-based assessment, a mutagenicity screening assay is usually conducted prior to selecting a lead compound. Some companies also perform screening assays for other genotoxicity endpoints such as DNA or chromosome damage, or change in chromosome number. Not all companies conduct these additional screening assays because as long as the drug is not a direct acting mutagen, it may still be developed if no genotoxicity is observed in vivo or as long as a sufficiently large margin is achieved between the genotoxic and clinical doses/exposures.

3.4.1 Structural Assessment

Computational structural assessments do not require actual drug synthesis and hence are usually the first safety evaluation conducted for drug candidates. Structural assessments may be conducted via literature review or by using commercially available software packages. There are several commercially available software programs that currently meet FDA basic requirements for transparency, predictive accuracy and other parameters. Refer to Chap. 2 by Stavistkaya, Aubrecht, and Khrulak for an in depth discussion on this topic.

3.4.2 Bacterial Mutagenicity Screening

Compounds that contain structural alerts for mutagenicity are further evaluated in a screening bacterial mutagenicity test to confirm the computational prediction. Prior to full development, only limited quantities of drug are generally available for early efficacy and safety studies. For these reasons genotoxicity testing using the regulatory GLP assay protocols are not normally conducted. In response to this limitation in drug supply, protocols have been developed that use small amounts of drug (mg quantities) but are sufficiently robust to identify the vast majority of mutagenic drugs before extensive resources have been expended on larger scale chemical synthesis and other drug development activities. Some tests are limited versions of the regulatory GLP test, detecting the exact same endpoint, while using fewer than the GLP required 5-bacterial strains, modified strains or entirely different bacterial strains, fewer replicate plates, smaller plates, lower top doses, colorimetric read-out rather than counting revertant colonies or a surrogate endpoint other than mutation. In this sense, the mutagenicity screening tests can be viewed as more "high-throughput" assays compared to the standard GLP study while maintaining high sensitivity with moderate to significant reductions in compound requirement. Table 3.2 lists a few of the most widely used bacterial screening assays comparing the amount of compound required for each and the strengths and weaknesses [1]. The performance of each of these assays was evaluated using a different number of reference compounds and different reference compounds, so the ability of each assay to identify mutagens cannot be directly compared. However, the performance of most of these assays can be increased if consideration is given to the results of the structural alert assessment. Several of the software systems provide detailed information regarding the strain or metabolic conditions associated with the mutagenic prediction. This information can be used to design an intelligent screening assay tailoring the strain or assay conditions to enable confirmation or de-risking a mutagenic structural alert.

For example, boronic acids and esters are a chemical class recently recognized to have mutagenic potential and such knowledge has been incorporated into recent mainstream computational software. Structural alerts and experience has shown that these compounds are often uniquely positive in E. coli or TA1535, strains not typically included in routine screening unless computational prediction is considered.

A second example, alkyl halides, have long been recognized to have mutagenic activity. Experienced researchers know that many compounds in this class are mutagenic in strains TA1535 and less potently in TA100. Use of structural alert software and knowledge based assessments can provide this information quickly without requiring extensive mutagenicity testing or literature searches.

3.5 GLP Battery Selection Strategy (Option 1 vs. Option 2)

Several factors should be considered when choosing between ICH S2(R1) Option 1 and Option 2 battery. Option 1 is based on the previously accepted testing battery described in ICH S2A/B. This option is preferred when an Ames assay may not be appropriate for a specific test article (e.g., antibacterials) or when non-clinical systemic exposure provides insufficient coverage relative to clinical systemic exposure. Although Option 2 has not previously existed as a stand-alone testing battery, the

Table 3.2 Bacterial screening assays

Assay	Plate/dish	Strains	Compound (mg)	Strengths	Limitations
Exploratory Ames	100 mm	TA98/TA100 at minimum & variable 3rd strain	125	Same method as GLP assay. Vary 3rd strain based upon structural alert information	May miss unique single strain positive compounds if no alert
MicroAmes (pre-incubation)	100 mm	TA98/100/1535 1537, E.coli	50	All 5 standard strains evaluated	Increased test article concentration may lead to increased precipitation interference
Mini-Ames Mix	100 mm	TA100 & mixed TA98, 1535, 1537	40	Standard strains	Missed weak and E.coli unique mutagens
Miniscreen	6-well	TA98/TA100 at minimum & variable 3rd strain	25	Standard strains	Same as exploratory Ames; Low spontaneous colony number
Micro-Ames	24-well	TA98,100,1535 TA97a	8	Standard strains	Low spontaneous colony number
BioLum Ames	24-well	TA98/100/1535 1537, E.coli [*lux*(*CDABE*) modified]	10	Accurate count of very small colonies	Modified bacterial strains; Low spontaneous colony number
5-Fluocuracil mutation assay	384-well	FU100 (modified TA100)	30	Fast colorimetric assay	Forward mutation resistance to 5-fluorouracil; low sensitivity to non-classical mutagenic alerts
Ames-II microtiter	384-well	TA98/TAmix (modified strains)	35	Fast colorimetric assay	Modified bacterial strains; insensitive to some chemical classes
SOS -response systems	96-well	PQ37 (modified E.coli)	2	Fast colorimetric assay	Modified strain; Reporter gene assay in-directly measuring mutation

approach is consistent with follow-up tests recommended for de-risking in vitro mammalian cell positives. This Option may be preferred in cases where short-lived reactive metabolites are expected to form in the liver or if a non-GLP screening study indicates a positive result would be expected in the GLP in vitro mammalian cell assay. Under these situations consider using Option 2 with evaluation of both genotoxicity endpoints (micronucleus and Comet) integrated into a single in vivo study to reduce animal use and maximize data acquired.

3.6 High-Dose Selection

In order to adequately characterize genotoxic potential for both in vitro and in vivo assays, it is important to test to appropriately high doses. This dose must be sufficiently high to detect relevant genotoxic responses but not overly high as several well-understood phenomena leading to irrelevant positives can occur. Important considerations for in vitro assays include test article solubility and cytotoxicity. The selection of an appropriate limit dose for in vivo assays also factors in the duration of dosing and whether genetic toxicology endpoints are combined with a general toxicology study. Specific details, as described in ICH S2(R1) are summarized below.

3.6.1 Ames Assay

For readily soluble and non-cytotoxic test articles, the maximum recommended dose for bacterial reverse mutation assay is 5000 μg/plate. When limits of solubility are exceeded, the lowest dose yielding precipitation should be scored as the top dose assuming no cytotoxicity is encountered. The top dose may also be limited to a level providing a significant degree of cytotoxicity (e.g., reduced background revertants or background lawn).

3.6.2 In Vitro Mammalian Cell Assays

The maximum recommended concentration is the lower of 1 mM or 0.5 mg/mL when solubility and cytotoxicity are not limiting. For insoluble but non-cytotoxic test articles, the top concentration should produce minimal precipitation as long as scoring is still possible. For cytotoxic test articles, top concentrations should need not exceed a ~50 % reduction in cell growth for chromosomal aberration or ~55 % reduction for micronucleus assays. When cytotoxicity is used to limit concentrations evaluated in the mouse lymphoma assay, the relative total growth should be ~10–20 %.

3.6.3 *In Vivo Assays*

The maximum recommended dose for acute studies (e.g., one to three dose administrations) is 2000 mg/kg/day or a dose producing the maximum tolerated dose (MTD). For testing conducted with in vitro mammalian assay positive test articles or when Option 2 is utilized, selection of a top dose must take into account several factors. These include the MTD (e.g., based on consideration of lethality following acute administration), maximum feasible dose (MFD) based on solubility, limit dose of 1000 mg/kg/day for studies ≥14 days, saturation of systemic exposure, and target tissue toxicity. In cases where the in vitro mammalian cell assay is negative and Option 1 is being followed, evaluation of a genetic toxicology endpoint can be combined with a standard general toxicology study. In such cases, it is reasonable for top dose selection to support clinical development as described in ICH M3(R2).

3.7 Follow-Up Testing for Ames Positives

Unless gene mutations are deemed an acceptable liability (e.g., acute treatment, life-threatening indication, etc.), it is often difficult to develop drugs that are positive in the Ames assay. The exception is positive results due to experimental artifact or those with questionable clinical relevance. Examples of artifact include the presence of histidine or tryptophan [2–5]. Positive results considered to be of questionable relevance are those related to mutagenic impurities that are not present in the clinical batch or formation of mutagenic metabolites through a rodent S9 or bacterial specific pathway (e.g., nitroreductase).

For relevant Ames positive drugs, there is no clear follow-up testing strategy. ICH S2(R1) simply indicates that "extensive follow-up testing" is needed to further characterize potential in vivo mutagenicity and carcinogenicity. Currently there is no single assay deemed sufficient to mitigate concerns with an Ames positive finding. However, there are multiple assays to consider as potential components of a follow-up testing strategy. For instance, evaluating the formation of large colonies in the mouse lymphoma assay provides a rapid in vitro assessment of gene mutations in mammalian cells. Two in vivo options are available to directly assess gene mutation events. One option is the transgenic gene mutation assays which is most often used as a tool to evaluate mechanism of action (e.g., establish potential of carcinogenic compounds to induce gene mutations as follow-up to a 2 year rodent bioassays). In addition, this assay can also be used to further characterize the in vivo relevance of an Ames positive finding. This assay allows evaluation of mutations in multiple tissues but requires 28 days of test article administration (specific models discussed in more detail in OECD TG488) [6]. Another recently developed in vivo option for assessing Ames positive findings is the Pig-a gene mutation assay (covered in a special issue of EMM) [7]. The Pig-a mutation is detected only in reticulocytes so tissue coverage is limited. A key advantage of this assay is the ability to integrate the endpoint in general toxicology studies using standard animal models. Pig-a mutations can also be measured in clinical samples. Additional in vivo options include evaluation of primary DNA damage in the UDS and comet assays as well as measurement of DNA adduct formation. These endpoints are potentially related to but do not directly measure gene mutations. The comet assay and measurement of DNA adducts can be applied to many tissues while UDS is restricted to the liver. UDS is generally considered to be an insensitive endpoint, with the potential exception of drugs that induce bulky DNA adduct formation. DNA adducts can be measured using relatively rapid and sensitive techniques, such as liquid chromatography – mass spectrometry based methods, but are not often submitted in support of regulatory decision making. Overall, the utility of the primary DNA damage assays is limited but perhaps useful for contributing to the weight of evidence against reactive potential.

In some cases, data from a short-term transgenic mouse carcinogenicity study (e.g., p53+/– or rasH2) has been requested by regulators. While providing definitive data on the endpoint of clinical concern (i.e., carcinogenicity), the studies require large number of transgenic animals and 6 months of dosing.

Ultimately, the ability to proceed with clinical administration of an Ames positive molecule depends on the strength of evidence demonstrating an acceptable risk:benefit profile.

3.8 Follow-Up Testing for In Vitro Mammalian Cell Positives

In vitro mammalian cell assay positives are more prevalent during drug development than bacterial mutation positives and are far less likely to prevent continued drug development. The rate of bacterial mutagens encountered is only ~10–12 % during lead compound identification, if synthetic intermediates and reactive synthetic reagents are discounted, while the rate of in vitro mammalian cell positives varies across experimental systems and can range as high as 25–30 % [8]. However, in many cases the positive in vitro response may not be relevant to safety at human exposure levels.

Under option 1 in ICH S2(R1), drugs should be evaluated for genotoxicity potential in mammalian cells in vitro . The sponsor has the choice of conducting an in vitro metaphase chromosome aberration assay (CAA), an in vitro micronucleus assay (IVTMN) or a mouse lymphoma L5178Y cell $TK^{+/-}$ (thymidine kinase) gene mutation assay (MLA). Selection of which in vitro mammalian cell assay to conduct is a decision each sponsor must make. Things to consider when selecting which assay to conduct is what endpoints are detected in each system, and for those conducting these studies in-house, equipment and expertise of those evaluating the data. CAA detects structural chromosome damage and numerical aberrations (i.e. polyploidy). IVTMN detects structural chromosome damage in addition to aneuploidy and qualitatively cell cycle perturbations, when evaluated using flow cytometry. MLA detects both point mutations and chromosomal mutations based upon the ratio of large and small mutant colonies. Time and cost required to conduct the CAA and MLA historically have been similar. The IVTMN test option was recently added as an option with the 2012 update of ICH S2(R1). All three assays (CAB, IVTMN and MLA) may be evaluated using manual microscopic methods but the IVTMN assay has the option of using automated flow cytometric evaluation which can reduce study duration to 1 week compared with 4–5 weeks for CAA and MLA.

Several approaches can be taken to investigate the relevance of in vitro positives including defining the mode of action (MOA) or accumulating more data and generating a weight-of-evidence (WOE) argument. Because defining an acceptable MOA argument is more difficult and potentially more time consuming, usually a WOE approach is taken within industry. A first step for any investigation should be to consider the reproducibility of the initial positive response and to consider repeating the test using a different cell line to rule out cell line specifics such as p53 status. A positive response should be reproducible and fall outside of the laboratory historical control range. Also, since pharmaceuticals are designed to have a pharmacological effect, consider whether the positive in vitro response could be due to exaggerated pharmacology.

3.8.1 Irrelevant In Vitro Positives

Irrelevant positive in vitro responses can generally be divided into three categories: (a) activity specific to in vitro culture; (b) non-DNA interactions with components of critical cellular functions; and (c) direct DNA damage at concentrations above a threshold due to disruption of basic cellular homeostasis.

The first step to evaluating any positive experimental data is to consider whether the test substance was sufficiently pure and stable, or whether an impurity or degradant could be responsible for the positive response. A broad spectrum of cell culture artifacts such as pH, osmolality, and excessive toxicity have long been recognized as sources of irrelative positive results [9]. In recognition that excessive toxicity could induce an irrelevant positive response, the top dose required for in vitro assays was reduced tenfold in the latest ICH S2(R1) guidance document.

Dimethyl sulfoxide (DMSO) is a commonly used solvent for in vitro assays and there have been occasions when the test substance has reacted with DMSO to generate a substance reported in the literature to be genotoxic. Before conducting additional in vitro studies consider consulting with a chemist on solvent selection to ensure that the test substance does not react with the selected solvent.

The cell line and origin of the cells being used to conduct genotoxicity testing should be considered and periodically re-evaluated. Over time due to poor cell culture, storage or labeling, spurious laboratory results may be attributed to instability of the cell line being used. Some labs have even found that cell lines thought to be of human origin were contaminated with rodent cells at some point. To guard against this as well as simple genetic drift ensure that target cells are performing within published parameters. The HESI Genetic Toxicology Technical Committee is in the process of establishing repositories of characterized L5178Y, TK6, CHO-WBL,

CHL, and Hep G2 cells for genotoxicity testing [10]. The repositories will be at Sigma (USA), ECACC (UK), and JCRB (Japan), and a guidance document on good cell culture is in preparation. It's important to verify that cell lines intended for genotoxicity studies exhibit appropriate background response frequencies, and respond to reference chemicals appropriately.

Cell culture systems are routinely supplemented with S9 liver metabolic fraction to mimic in vivo metabolic activation because the most commonly used immortalized cell lines contain little to no metabolic activity. An exception to this are cell lines genetically engineered to express p450 activity or immortalized hepatocytes. The S9 liver metabolic fraction typically used to supplement in vitro genotoxicity studies originates from rats induced to have high CYP P450 levels by pre-treatment with Aroclor 1254 or phenobarbital/β-napthoflavone prior to preparation of S9. The metabolite profile produced using induced S9 in vitro can be very different than the metabolite profile in un-induced rats or humans. In vitro only phase 1 metabolism is present because the S9 system is only supplemented with CYP co-factor NADPH not co-factors required for phase 2 metabolism. Therefore, a genotoxic metabolite inducing a positive in vitro response, may not be relevant in vivo because it may be detoxified by conjugation or may be produced at very low levels due to competing or alternative CYP pathways dominating in vivo.

Consideration should also be given as to whether the genotoxic effect can be attributed to a non-DNA reactive mechanism. Sometimes test substances do not directly interact with DNA but rather target macromolecules involved in critical cellular functions such as cell division [11–13]. Test substances which bind to elements of the mitotic apparatus generally result in aneuploidy and are easily detected in the IVMN assay and to a lesser extent in the CAA. Examples of non-DNA reactive in vitro cell assay positives have been demonstrated for test substances that inhibit DNA synthesis and repair enzymes, perturb nucleotide pool balance, or generation of reactive oxygen species via lipid peroxidation reaction and others have been described by Scott et al. [9] and Kirkland et al. [14].

Consideration should be given as to whether the positive response is due to direct DNA interaction or only observed above a defined threshold. A common example of this are kinase inhibitors designed to have selectivity at therapeutic exposures that maybe in the nanomolar range, but concentrations required for in vitro genotoxicity testing may be >10,000-fold higher than intended therapeutic doses. At such high concentrations pharmacologic selectivity is lost and the effect is non-specific inhibition of kinases including those kinases responsible for maintaining cellular homeostasis. As a class many kinase inhibitors are positive in CAA and IVTMN assays.

3.8.2 Selection of Appropriate Follow-Up Tests

Follow-up testing strategy is dependent upon the genotoxicity assay in which the positive response was observed and has been the basis of several industry discussions and publications [14–16]. For any of the in vitro assays where the positive result is seen only in the presence of S9 metabolic activation, consideration should be given to comparing the in vitro metabolic profiles from rats and humans. Consideration should also be given as to whether the lack of phase 2 conjugation may be contributing to the positive response. The in vitro study could be repeated with addition of phase 2 co-factors to investigate this possibility. For MLA, selection of follow-up tests depends upon the proportion of small and large colonies. If the positive response was due to predominately small colony mutants then an appropriate follow-up in vitro assay would be one that detects clastogenicity such as CAA or IVTMN. However if large colonies were predominately responsible for the positive response then consideration should be given to an *hprt* assay or the in vitro comet assay. Positive MLA small colony results would add WOE that a positive CAA response was relevant, because both in vitro assays detect the same chromosome mutation

Generation of WOE information using additional in vitro studies helps clarify the relevance of the initial positive response, but moving directly to an in vivo study may be a faster and more reliable approach to reducing perceived risk. ICH S2(R1) requires negative results for 2 in vivo genotoxicity endpoints to conclude that an in vitro positive response is not biologically relevant. The guidance recommends conducting a comet assay evaluating DNA damage in liver, and potentially known target tissues and a micronucleus assay evaluating induction of chromosome damage in bone marrow or peripheral blood. While a micronucleus evaluation is easily integrated into general toxicity studies, the comet assay cannot be easily integrated due to logistical hurdles arising from the short 3–6 h exposure condition. Whenever a Comet assay is needed, a standalone in vivo genotoxicity study including Comet and micronucleus assays is often the best solution.

For test substances found to be genotoxic in vitro but not in vivo, there should be evidence that the test substance reached the target organ in vivo. This can be addressed for negative micronucleus studies by always measuring test substance exposure in the blood, because the bone marrow is a highly perfused organ. For orally dosed comet assays, site of contact tissues do not require proof of exposure and similar to bone marrow, liver is a highly perfused tissue and generally demonstrating systemic exposure is sufficient proof of tissue exposure.

Consider the in vitro endpoint (mutagenicity, clastogenicity, or aneugenicity) when selecting in vitro or in vivo follow-up tests. Ensure that the follow-up test selected can detect the endpoint that needs to be verified. For test substances inducing only gene mutations in vitro, a transgenic mouse gene mutation assay or rodent Comet assay would be appropriate follow-up studies. These assays have OECD test guidelines describing proper assay conduct. Other follow-up assays could be used with scientific justification even those without formal OECD test guidelines, but care should be taken to design these studies in accordance with scientific literature. For test substances only inducing chromosomal aberrations, appropriate follow-up in vitro tests would include chromosome aberrations in an alternative cell line, micronucleus or comet assay, and an in vivo micronucleus or Comet test. For test substances inducing both gene mutation and clastogenicity in vitro consider an in vivo Comet and micronucleus assay to verify in vitro findings. Sponsors could consider conducting an in vivo mutation endpoint such as a transgenic mouse or Pig-a gene mutation assay to further investigate potential in vivo mutation. For test compounds only inducing genotoxicity in the absence of S9 metabolic activation, consideration should be given to conducting a Comet assay including a site of contact tissue (eg. GI tract, or skin depending upon dosing route) in addition to liver.

3.9 Case Studies

3.9.1 Positive Ames – Peptide

A peptide containing histidine and tryptophan amino acids tested positive in a plate incorporation bacterial mutagenicity test. The positive response was weak, only exceeding the 2.0-fold positive response criteria at the highest dose evaluated. This type of response could have been due to an impurity in the test substance or due to release of histidine from the peptide. Re-purification of the peptide would be labor intensive, so a "feeding effect" from histidine was evaluated first. The Ames test was repeated using the method where the bacteria are pre-incubated with the test substance prior to plating on agar plates. This provided the opportunity to wash the bacteria several times to remove any residual peptide (histidine/tryptophan source) prior to plating the bacteria on agar(2). The pre-incubation method was negative with no increase in revertant colony count and no increase in toxicity. Based on this information the peptide was considered not mutagenic. The small increase in revertant colony counts was attributed to facilitation of growth of auxotrophic colonies by the additional histidine present in the treated plates, but not the vehicle plates.

3.9.2 Genotoxic In Vitro but Non-genotoxic In Vivo

An impurity was mutagenic in the Ames test (four-fold maximum response) and clastogenic in the in vitro micronucleus assay in CHO cells (six-fold maximum response) in the absence of S9 metabolic activation. The genotoxicity response was further evaluated in vivo to provide data supporting the hypothesis that the genotoxic impurity quickly degraded to a non-genotoxic species in the acid pH of the stomach. Rats were orally administered the test substance daily for 1-month with strong systemic exposure to the test substance measured on Day 1 and during Week 4, with Cmax concentrations in vivo equal to or greater than in vitro concentrations where genotoxicity was observed. However, the test substance did not induce a mutant phenotype in the Pig-a gene mutation assay sampled on Day 31, did not induce micronuclei in peripheral blood sampled on Day 14, and did not induce DNA damage (comet assay) in duodenum or liver following 3–6 h or 24-h exposure.

Based on the weight of evidence it was concluded that the test substance was not genotoxic in vivo.

The Pig-a erythrocyte mutation assay has been demonstrated to accurately detect mutagenic test substances. The Pig-a assay was selected to evaluate in vivo mutagenic potential rather than one of the lacZ reporter gene based transgenic assays because the Pig-a assay could be integrated into a 1-month rat study using the same strain animal as previous toxicology studies enabling leverage of all information previously gathered for this test substance. Also the Pig-a assay is much cheaper and faster than the lacZ based transgenic assays or the transgenic carcinogenicity models (eg. $p53^{+/-}$, or rasH2).

3.9.3 Positive In Vitro Chromosome Aberrations with S9 Metabolic Activation Only

A test substance induced structural chromosome aberrations only in the presence of S9 liver metabolic activation from Aroclor induced rats. The metabolite profile generated in the tissue culture media with test substance and S9 but without cells was evaluated and it was postulated that the major metabolite generated in the in vitro S9 system would not accumulate or persist due to rapid detoxification (glucuronidation). The chromosome aberrations assay with S9 metabolic activation only was repeated as before but with the addition of a secondary set of cultures containing glutathione at physiological concentration. The positive response with S9 was reproduced, but the positive response was completely ameliorated in the glutathione containing cultures. Based upon these results, in addition to data demonstrating the test substance did not induce micronuclei in rats, the genotoxic response in the in vitro chromosome aberrations assay was considered biologically irrelevant with no impact on potential patient safety.

3.9.4 Positive In Vitro Micronucleus – Aneugen

A test substance induced a significant increase in micronuclei in vitro in CHO cells both in the presence and absence of S9 liver metabolic activation. A micronucleus assessment was piggybacked on an on-going 2-week rat general toxicology study to evaluate the biological relevance of the positive response. Micronucleus induction was evaluated in peripheral blood collected on Day 7 and rapidly evaluated using flow cytometry. The test substance did not induce micronuclei in rats when tested up to maximum limits. Cmax data indicated that systemic exposures achieved in rats were ~50 % lower than the in vitro concentration where micronuclei were significantly induced. Drug development of the candidate continued.

Another test substance also induced significant increase in micronuclei in vitro in CHO cells and an in vivo micronucleus assessment was piggybacked on a 1-month rat general toxicology study. However, in this case the test substance induced micronuclei in rats at doses that resulted in systemic exposure multiples of 300× the predicted human exposure at the highest anticipated clinical dose. Based on the exposure multiple the test substance continued in development.

A test substance was demonstrated to induce micronuclei in vitro, but based upon a meaningful increase in the hypodiploidy gate observed during flow cytometric evaluation, the response was probably due to aneuploidy (chromosome loss). An aneuploidy response can occur when the test substance interferes with the mitotic apparatus leading to non-disjunction resulting in a micronucleus that contains an entire chromosome. Aneuploidy is widely accepted as an example of a threshold response. Therefore, a second in vitro micronucleus study was conducted to determine whether the origin of the micronuclei was from aneuploidy (kinetochore positive micronuclei) or an acentric chromosome break (kinetochore negative micronuclei). The in vitro kinetochore study demonstrated that >85 % of the micronuclei induced were due to an in-direct aneuploidy mechanism and not direct DNA damage. Therefore it was concluded that the test substance was safe for clinical trial volunteers at exposures less than those that induced aneuploidy.

3.9.5 Positive In Vivo Micronucleus with Elevated Body Temperature

A test substance was found to induce micronuclei formation following acute dosing in rats. The effect was only observed at a dose known to also cause an increase in body temperature in rats. A second in vivo micronucleus assay was performed using a repeat-dose protocol with doses that were not associated with temperature changes. Results from this assay were negative at systemic exposures providing a robust safety margin vs. expected clinical exposures. The sponsor also conducted an in vivo UDS assay in liver to demonstrate a lack of DNA reactivity. The results from this study were also negative. Although there is minimal regulatory confidence in the UDS assay, the results contributed to the weight of evidence argument. Based on the totality of information available, the positive acute in vivo micronucleus results were determined to be the result of an indirect mechanism. The safety margins established in the repeat-dose micronucleus assay were deemed sufficient to allow clinical development to proceed.

Both hyperthermia and hypothermia have been demonstrated to induce micronuclei via an indirect mechanism of action in both mice and rats, so the potential impact of body temperature changes should be considered when evaluating positive in vivo micronucleus results [17–20].

References

1. Escobar PA, Kemper RA, Tarca J, Nicolette J, Kenyon M, Glowienke S et al (2013) Bacterial mutagenicity screening in the pharmaceutical industry. Mutat Res 752(2):99–118
2. Khandoudi N, Porte P, Chtourou S, Nesslany F, Marzin D, Le Curieux F (2009) The presence of arginine may be a source of false positive results in the Ames test. Mutat Res 679(1–2):65–71
3. Thompson C, Morley P, Kirkland D, Proudlock R (2005) Modified bacterial mutation test procedures for evaluation of peptides and amino acid-containing material. Mutagenesis 20(5):345–350
4. Glatt H, Oesch F (1985) Mutagenicity of cysteine and penicillamine and its enantiomeric selectivity. Biochem Pharmacol 34(20):3725–3728
5. Aeschbacher HU, Finot PA, Wolleb U (1983) Interactions of histidine-containing test substances and extraction methods with the Ames mutagenicity test. Mutat Res 113(2):103–116
6. OECD (2013) Test No. 488: Transgenic rodent somatic and germ cell gene mutation assays, OECD guidelines for the testing of chemicals, section 4. OECD Publishing, Paris. doi:http://dx.doi.org/10.1787/9789264203907-en
7. Special issue of environmental molecular & mutagenesis focused on the Pig-a gene mutation assay (2011) 52(9):681–794
8. Kirkland D, Pfuhler S, Tweats D, Aardema M, Corvi R, Darroudi F, Elhajouji A, Glatt H, Hastwell P, Hayashi M, Kasper P, Kirchner S, Lynch A, Marzin D, Maurici D, Meunier J-R, Muller L, Nohynek G, Parry J, Parry E, Thybaud V, Tice R, van Benthem J, Vanparys P, White P (2007) How to reduce false positive results when undertaking in vitro genotoxicity testing and thus avoid unnecessary follow-up animal tests: Report of an ECVAM Workshop. Mutat Res 628:31–55
9. Scott D, Galloway SM, Marshall RR, Ishidate M, Brusick D, Ashby J, Myhr BC (1991) Genotoxicity under extreme culture conditions. A report from ICPEMC task group 9. Mutat Res 257:147–204
10. Lorge E, Moore MM, Clements J, O'Donovan M, Fellows M, Honma M, Kohara A, Galloway S, Armstrong MJ, Sutter A, Thybaud V, Gollapudi B, Aardema MJ, Tanir JY (2015) Standardized cell sources and recommendations for good cell culture practices in genotoxicity testing. Mutat Res (in preparation)
11. Parry JM, Parry EM, Boumer R, Doherty A, Ellard S, O'Donovan J et al (1996) The detection and evaluation of aneugenic chemicals. Mutat Res 353(1–2):11–46
12. Schuler M, Muehlbauer P, Guzzie P, Eastmond DA (1999) Noscapine hydrochloride disrupts the mitotic spindle in mammalian cells and induces aneuploidy as well as polyploidy in cultured human lymphocytes. Mutagenesis 14(1):51–56
13. Rao CV, Yamada HY, Yao Y, Dai W (2009) Enhanced genomic instabilities caused by deregulated microtubule dynamics and chromosome segregation: a perspective from genetic studies in mice. Carcinogenesis 30(9):1469–1474
14. Kirkland DJ, Aardema M, Banduhn N, Carmichael P, Fautz R, Meunier J-R, Pfuhler S (2007) In vitro approaches to develop weight of evidence (WoE) and mode of action (MoA) discussions with positive in vitro genotoxicity results. Mutagenesis 22:161–175
15. Thybaud V, Aardema M, Casciano D, Dellarco V, Embry MR, Gollapudi BB, Hayashi M, Holsapple MP, Jacobson-Kram D, Kasper P, MacGregor JT, Rees R (2007) Relevance and follow-up of positive results in in vitro genetic toxicity assays: an ILSI-HESI initiative. Mutat Res 633(2):67–79
16. Dearfield KL, Thybaud V, Cimino MC, Custer L, Czich A, Harvey JS, Hester S, Kim JH, Kirkland D, Levy DD, Lorge E, Moore MM, Ouedraogo-Arras G, Schuler M, Suter W, Sweder K, Tarlo K, van Benthem J, van Goethem F, Witt KL (2011) Follow-up actions from positive results of in vitro genetic toxicity testing. Environ Mol Mutagen 52(3):177–204
17. Asanami S, Shimono K (1997) High body temperature induces micronuclei in mouse bone marrow. Mutat Res 390(1–2):79–83

18. Asanami S, Shimono K (1997) Hypothermia induces micronuclei in mouse bone marrow cells. Mutat Res 393(1–2):91–98
19. Asanami S, Shimono K (2000) Effects of chemically- and environmentally-induced hypothermia on micronucleus induction in rats. Mutat Res 471(1–2):81–86
20. Asanami S, Shimono K (2009) Species-level differences between mice and rats in regards to micronucleus induction with the hypothermia-inducing drug haloperidol. Mutat Res 676(1–2):102–105

Chapter 4
Genotoxic Impurities in Pharmaceuticals

Peter Kasper and Lutz Müller

Abstract Reliable quality is one of the key attributes of drugs nowadays. Patients deserve the highest quality and are expecting to not be put at risk for health effects especially related to impurities in drug substances or drug products. While ICH guidelines for "ordinary" impurities have been available for many years, a harmonized guideline on how to assess, limit and control potential health effects of low levels of genotoxic/carcinogenic impurities was lacking and only regional (draft) guidelines existed. With the ICH M7 guideline entitled "Assessment and control of DNA-reactive (mutagenic) impurities in pharmaceuticals to limit potential carcinogenic risk", this gap in the internationally harmonized regulatory framework has been filled. This chapter deals with the history of the guideline and concepts of risk assessment with a focus on the Threshold of Toxicological Concern (TTC) principle. Further, a few examples are given on how to deal with potential impurities with mutagenic/carcinogenic potential or for compounds for such a potential is assumed but not demonstrated. Hence, in the following chapter, the reader can expect some background information about the ICH M7 guideline and tips how to use it in practice. Regulatory precedence of the use of evidence of non-linear dose–response for genotoxic carcinogens such as EMS is also referred to.

Keywords Ames test • Cancer risk assessment • DNA-reactive impurities • Genotoxic impurities • Impurities in pharmaceuticals • In silico assessment • QSAR • Threshold of toxicological concern • TTC

P. Kasper (✉)
Genetic and Reproductive Toxicology, Federal Institute for Drugs and Medical Devices (BfArM), Bonn, Germany
e-mail: Peter.Kasper@bfarm.de

L. Müller
F. Hoffmann La-Roche, Pharmaceutical Sciences, Roche Innovation Center Basel, Bldg. 73/2b, Basel CH-4070, Switzerland

M.J. Graziano, D. Jacobson-Kram (eds.), *Genotoxicity and Carcinogenicity Testing of Pharmaceuticals*, DOI 10.1007/978-3-319-22084-0_4

4.1 Introduction

To date, most drugs to treat human disease are "small molecule" chemical entities. The selection of the so-called "active pharmacological ingredient" or API for clinical development is based on a series of non-clinical and clinical pharmacology, ADME and toxicity/safety studies. The API is mostly chemically synthesized using processes to comply with resource need, cost of goods, yield in the synthetic steps and feasibility to eventually obtain a highly pure chemical, which drives the efficacy and safety of the drug. As these small molecules tend to be highly complex chemicals carrying various functional groups to confer target interaction, to protect from certain metabolism pathways, to enable bioavailability and excretion, the synthesis most often involves many steps. Like for any other chemical, the term "purity" is to be used in the context of understanding limitations of such processes. Hence, any small molecule API will always carry with it process-related material, i.e. "impurities". These could be a safety concern, especially if their toxicity could not be monitored in humans e.g. if they were mutagenic carcinogens. Similarly, an API will usually be administered in a galenic form with inclusion of excipients, fillers, colorants, stabilizers, etc. This galenic form brings in additional impurities as well as the potential for generation of reaction products with the API. These mostly build up under storage conditions. These two conditions already determine major differences between impurities in the API, so-called "drug substance impurities" and impurities in the galenic form, so-called "drug product impurities". Whereas drug substance impurities can be managed by the pathways and ingredients used in the synthesis process (and can sometimes be avoided if necessary), drug product impurities most often are controlled by packaging or storage conditions or shelf-life. They usually cannot be avoided. This is important especially when safety considerations are of very high stringency, i.e. for unusually potent or toxic impurities.

Under a regulatory point of view control of impurities in the drug substance and degradants in APIs are addressed in the International Conference of Harmonization Quality Guidelines Q3A(R2) [1] and Q3B(R2) [2], respectively, and the Q3C(R5) guideline [3] that deals with residual solvents. However, these documents do not provide any specific guidance for determining acceptable levels for genotoxic impurities. In these guidelines, one sentence elaborates on identification and control of unusually toxic impurities as follows: "For impurities known to be unusually potent or to produce toxic or unexpected pharmacological effects, the quantitation/detection limit of the analytical procedures should be commensurate with the level at which the impurities should be controlled." This sentence could be interpreted in a way that e.g. immunogenic, carcinogenic and genotoxic impurities would potentially need lower specification limits than given by these guidelines for ordinary impurities. However, the guidelines do not give any specifics how to do that in practice.

In this context, the process for highly toxic "Class 1" residual solvents as described in the ICH guideline Q3C is helpful. This guideline follows a strict policy of avoidance of such Class 1 compounds especially if these have genotoxic and carcinogenic properties. The genotoxic carcinogen benzene belongs to the solvents of class 1 and the guideline stipulates that such solvents "should not be employed in the manufacture of drug substances, excipients, and drug products because of their unacceptable toxicity". However, if its use is unavoidable "in order to produce a drug product with significant therapeutic advance" then the level of benzene should be restricted to 2 ppm. Such a concentration-based limit has severe shortcomings as it ignores large differences in human exposure whether the daily dose of a drug were 1 mg or 2 g without even considering the impact of duration of use, which is also ignored but is certainly a factor for assessing risks stemming from any exposure to a carcinogen like benzene. Hence, for limitations of risk based on total daily dose, a generic ppm limit does not make scientific sense. Yet, such generic limits allow establishment of generic analytical methods of detection and hence are seen useful for technical development due to their practicality and applicability across projects. If the generic level is derived from a worst case high daily dose intake scenario, it protects for all lower dose scenarios as well but is unnecessarily stringent.

Another example illustrating how regulatory control of drug impurities with a genotoxic and carcinogenic potential was addressed in the past is the EU 'Note for Guidance on Limitations to the Use of Ethylene Oxide in the Manufacture of Medicinal Products' [4]. Ethylene oxide is a highly reactive compound and is used in the synthesis of pharmaceutical raw materials and as a sterilant. According to IARC ethylene oxide is carcinogenic to humans (group 1) based on sufficient evidence from animal studies for carcinogenicity and compelling data in support of a genotoxic mode of action [5]. In view of this potential the EMEA Note for Guidance stipulates that the use of ethylene oxide in the manufacture of medicinal products "is acceptable only when pharmaceutically absolutely necessary, and then residual ethylene oxide in the product should not exceed a limit of 1 ppm" [4]. Interestingly, this limit is based on analytical feasibility rather than toxicological considerations; 1 ppm was considered the limit of detection for ethylene oxide residues at the time the Guideline was released.

4.2 The Emergence of a New Policy: Development of Genotoxic Impurity Guidelines

The European Medicines Agency (EMA) issued a "Guideline on the Limits of Genotoxic Impurities" in 2006 [6] in continuous evolution of a regulatory 'Position Paper' on the same subject that was published for comments 5 years earlier [7]. The intention of these documents was to fill in the gap identified in the existing ICH Q3 impurity guidelines when it comes to the question of up to what level an impurity would be acceptable in a pharmaceutical product when it is identified as a mutagen (referred to in the ICH Q3A guideline as impurity of "unusual toxicity"). In recognizing the continuing improvement in analytical chemistry capabilities and the possibility of detection of very low residues of impurities, a policy of a zero-risk and strict avoidance of potentially mutagenic impurities has been considered as an unrealistic option. The EMA guideline therefore proposed a health-protective daily intake level of 1.5 μg for genotoxic impurities based on the principles of a threshold of toxicological concern (TTC, for details see below). From this TTC value a concentration-based limit (ppm) can easily be calculated when the daily dose of the drug is known (rather than setting a fixed ppm-threshold for an impurity resulting in variable human intake levels as per the ICH Q3A/B approach). However, the 2007 EMA guideline still suggested that if avoidance was not possible, then residues of any genotoxic materials, even when below TTC, should be removed to a level that is "as low as reasonably practicable", the so-called "ALARP" principle. Due to lack of definition of what "reasonably" in this context could mean concern was expressed that this policy would lead to inconsistencies in setting acceptable intake levels based on ALARP considerations among regulatory authorities. These concerns were later addressed in a 2007 EMA Q&A-document which was amended in 2009 and 2010 [8]. The answer to Question #2 of the EMA document clarifies that "if the level of a mutagenic impurity is below the threshold of toxicological concern (equivalent to a clinical dose of ≤1.5 μg/day) it is not necessary to apply ALARP considerations" [8]. This is the very first regulatory document within the pharmaceutical sector that generally accepts the presence of mutagenic impurities in pharmaceuticals up to a certain level without the need for providing any specific justifications!

Another considerable shortcoming of the 2007 EMA guideline identified after its implementation was the lack of a modified TTC for shorter treatment durations. As a consequence, some regulatory authorities requested to apply the lifetime TTC value even for products used in early clinical development for short durations of treatment and in small populations. In 2006, a task force established under the umbrella of the US Pharmaceutical Research and Manufacturing Association (PhRMA) for the first time proposed the 'staged TTC' concept to be applied to pharmaceuticals [9]. The task force was established as a response to various clinical holds imposed by the FDA on investigational drugs in clinical trial phases based on suspicions that the drugs contained genotoxic impurities at levels potentially associated with a risk for the volunteers or patients involved in these trials. Thus, a staged approach to determine acceptable levels of genotoxic impurities for such situations and differentiating from the lifetime intake TTC was needed. The staged approach allows levels of daily intake of mutagenic impurities higher than 1.5 μg as defined by the lifetime TTC, namely 10 μg (for a 6–12 months duration), 20 μg (3–6 months), 40 μg (1–3 months), and 120 μg for not more than 1 month. The EMA adopted the staged TTC approach for limits of genotoxic impurities in clinical trials in the 2007 Q&A document [8], but to be more conservative it reduced the staged TTC limits proposed in the PhRMA paper by a factor of 2.

In 2008, the FDA issued a draft 'Guidance for Industry on Genotoxic and Carcinogenic Impurities in Drug Substances and Products: Recommended Approaches' [10] which was largely similar to the EU guidance. However, this document has not been finalized because in 2009 the topic 'genotoxic impurities' was adopted by ICH for development of a new internationally harmonized guideline.

Since the topic was considered to include both, safety and quality aspects the projected guideline was assigned to the M (multidisciplinary) series of the ICH process and designated as ICH M7 with the title "Assessment and Control of DNA-Reactive (Mutagenic) Impurities in Pharmaceuticals to Limit Potential Carcinogenic Risk". In February 2013 a draft of the M7 guideline was published in the three ICH regions for public consultation (step 3 of the ICH process). The document was adopted as a step 4 ICH Harmonised Tripartite Guideline in June 2014 [11] and is currently moving to the final step of the process that is the regulatory implementation.

4.3 Internationally Harmonized Regulations: The ICH M7 Guideline

The ICH M7 guideline is based in many aspects on the principles set by the EU- and the draft FDA guidelines on genotoxic impurities and thus the ICH M7 guideline does maintain continuity for industry and health authorities in this regard. However, some distinct differences and amendments can be recognized. In the following, general principles as well as practical recommendations of ICH M7 will be discussed in detail.

4.3.1 Genotoxic or DNA-Reactive Impurities: What Is the Difference?

The term 'genotoxic impurities' as applied by the EU Guideline on the Limits of Genotoxic Impurities has been replaced in ICH M7 by the term 'DNA-reactive impurities'. This change in terminology does not reflect a change of concepts since also in the EU guideline, genotoxic impurities are defined as "DNA-reactive substances that have the potential for direct DNA damage". However, using the term 'genotoxic' can be misleading as it includes mechanisms of genotoxicity where both, DNA-reactive as well as non-DNA-reactive compounds can be involved.

DNA-reactive compounds typically bind covalently to DNA resulting in DNA adducts which, if unrepaired can lead to mispairing during DNA synthesis thus giving rise to irreversible point mutations. Misrepaired DNA adducts can also result in strand-breakage and thus induce chromosome breaks. DNA-reactive compounds are therefore both, potential mutagens and clastogens. Non-DNA-reactive genotoxicants on the other hand target mainly components required for chromosome segregation (e.g. microtubules, kinetochores, centrioles), DNA synthesis (e.g. topoisomerases, DNA polymerases, imbalanced nucleotide pools) or other cellular enzymes (e.g. endonucleases, ligases), which subsequently give rise to structural or numerical chromosomal abnormalities but do not induce point mutations.

In more technical terms, a DNA-reactive chemical is usually detected as a mutagen in a bacterial reverse mutation (Ames) assay but could in addition also be positive for clastogenicity in test models for chromosomal damage. A non-DNA-reactive genotoxicant would typically be an Ames-negative compound with positive findings in chromosome damage tests only, such as cytogenetic assays or micronucleus tests.

The scientific basis for focusing on DNA-reactive rather than non-DNA-reactive molecules in the strategy of identifying genotoxic impurities of most concern is twofold. Firstly, DNA-reactive, Ames-positive compounds show a reasonably high correlation to rodent carcinogenicity whereas Ames-negative compounds with positive in vitro chromosomal aberration assay results have been shown to be poorly correlated with carcinogenic potential in rodents with more than 75 % noncarcinogens giving positive clastogenicity results [12]. Secondly, non-DNA-reactive genotoxic carcinogens typically have threshold mechanisms and usually do not pose carcinogenic risk in humans at the low level ordinarily present as impurities. In contrast, the default assumption widely applied in regulatory risk assessment for DNA-reactive genotoxic agents still is that they have linear dose responses without a threshold.

4.3.2 *Defining Acceptable Intake Level for DNA-Reactive Impurities: The TTC Concept*

Following the above assumption that DNA-reactive Ames-positive compounds are likely carcinogens with no threshold mechanism it would be theoretically impossible to define an absolute safe exposure, i.e., a zero risk level. Since it is commonly accepted that complete avoidance of small traces of DNA-reactive impurities in pharmaceuticals is often not a realistic option, implementation of a concept of acceptable risk is required. The threshold of toxicological concern (TTC) is potentially such a concept as it is based on the principle of establishing a human exposure threshold value below which there is a very low probability of an appreciable risk to human health. The TTC concept is designed to assess substances of which structural information is available but toxicological information is lacking. In the context of ICH M7 and the control of mutagenic impurities, structural information would refer to structural alerts for DNA reactivity causing a concern for carcinogenicity. If rodent carcinogenicity data are lacking in this situation a data-driven risk assessment to define an acceptable intake is unfeasible and a generic TTC value would be applied as an acceptable intake level that poses a negligible risk of carcinogenicity. However, if rodent carcinogenicity data are available for a (potentially) mutagenic impurity, application of the TTC concept is considered inappropriate and a compound-specific calculation of acceptable levels of impurity intake is recommended as is described in more detail below.

4.3.3 How Is the TTC for DNA-Reactive Carcinogens Derived?

The TTC concept was originally developed as a "threshold of regulation" at the FDA mainly to provide a practical approach for controlling traces of potentially carcinogenic components leached from food contact materials [13, 14]. It is based on an analysis of initially 343 chemical carcinogens from the Carcinogenic Potency Database (CPDB) [15]. The probability distribution of carcinogenic potencies expressed as TD50 values (the daily dose rate required to induce a calculated 50 % tumor incidence) has been used to derive an estimate of the dietary concentration of most carcinogens which would give rise to less than a one in a million (1×10^{-6}) upper bound lifetime risk of cancer. This very low risk level is generally recognized as a "virtually safe dose". In practice, the '10^{-6}-risk-equivalent' exposure distribution was achieved by simply dividing all TD50 values corresponding to a 0.5 risk level by 500,000. That dietary concentration was estimated to be 0.5 ppb (0.5 μg chemical/kg food), from which a human daily exposure level of 1.5 μg/person was derived, assuming that the whole amount of 1.5 μg is distributed throughout the total diet (1500 g of food, 1500 g of fluids). By expanding the database to more than 700 carcinogens the TTC value of 1.5 μg/person/day was repeatedly confirmed [16, 17]. In deriving the TTC value of 1.5 μg/day no distinction was made in the early analysis between genotoxic and non-genotoxic rodent carcinogens. However, further analysis of the CPDB clearly showed that the median carcinogenic potency of Ames-positive rodent carcinogens was about eightfold higher than the potency of carcinogens that are negative in the Ames test and it was therefore proposed to apply a tenfold lower TTC (0.15 μg/day) for chemicals with structural alerts that raise concern for potential mutagenicity [16, 18]. It is this TTC derived from a data set of Ames positive rodent carcinogens that is used as a point of reference in ICH M7 for impurities with a mutagenic potential. However, a lifetime risk of cancer of 1 in 10^5 (rather than 1 in 10^6) is considered acceptable for impurities in pharmaceuticals due to risk-benefit considerations and therefore the respective TTC value in this application is 1.5 μg/day for lifetime daily exposure. Excluded from the TTC are certain structural classes that were identified to be of such high cancer potency that intake even below the TTC may result in a significant carcinogenic risk. Mutagenic carcinogens belonging to this so-called cohort of concern are N-nitroso-, azoxy- and aflatoxin-like compounds [18].

It should be emphasized here that the concept to derive the TTC is based on a number of worst-case assumptions and therefore resulting in a highly conservative TTC value. For example, it is assumed

- that DNA-reactive and/or Ames positive compounds are human carcinogens; however, although the Ames test has good specificity (low frequency of false positives) for rodent carcinogens, the results of the rodent carcinogenicity studies are rarely corroborated by human correlates;
- that human risk at low doses may be estimated reliably by using a simple linear proportional extrapolation model on TD50 values (from the most sensitive site and the most sensitive species) derived from the rodent bioassays;

- That daily dosing is for lifetime i.e., 70 years in humans (corresponding to the 2-year lifetime duration of rodent carcinogenicity studies);
- and that the carcinogens used to create the CPDB, with many of high carcinogenic potencies, comprise a representative set of chemicals that could reasonably be expected to be components in pharmaceutical synthesis.

Therefore, rather than utilizing this very conservative generic TTC value in a 'one-fits-all'-approach to all pharmaceuticals, the built-in conservatism provides the opportunity to modify the TTC, for instance in relation to shorter treatment durations, availability of additional toxicological data and information about the clinical use/benefit of the drug (see below).

4.3.4 Hazard Identification: How Are Mutagenic Impurities Detected?

In order to minimize any safety risks stemming from toxic impurities in the final drug product, control of impurities is an important part of the drug manufacturing process. According to the recommendations of the ICH Q3A guideline an impurity would be considered qualified at the level present in the new drug substance batches used in the non-clinical safety studies including tests for genotoxicity. However, most impurities are usually present at levels below one percent of the drug substance and are therefore tested at doses/concentrations that would be in many cases below the No-Observed-Adverse-Effect-Level. When applying this qualification strategy many potentially mutagenic and carcinogenic impurities would therefore remain undetected and could thus be present in the pharmaceutical at levels that clearly exceed the theoretical acceptable cancer risk level. If, for instance, a batch of drug substance with an impurity at a level 0.1 % is tested in the Ames test the resulting maximum test concentration of the impurity is only 5 μg/plate, provided the API is tested up to the highest test concentration of 5000 μg/plate. In case of toxicity of the API to bacteria which would require a reduction of the top concentration the exposure to the impurity would even be lower. Based on a literature survey of approximately 450 mutagens (Ames positives) it was estimated that 85 % of mutagens are detected in the Ames test if the test concentrations goes up to at least 250 μg/plate [19]. On the other hand, 75 % of the mutagens would have been missed in the Ames test if the maximum concentration does not exceed 2.5 μg/plate. The latter would represent a common scenario of an impurity present at 0.2 % in an API testing batch used for bacterial mutagenicity testing at a reduced top concentrations of 1000 μg/plate due to toxicity resulting in a maximum test concentration of 2 μg/plate for the impurity.

Due to the recognized insensitivity of the ICH Q3A qualification approach in detecting mutagenic impurities in drug substances, alternative strategies are needed. Testing all impurities identified in a drug substance for mutagenicity using neat material is not a feasible option. Instead, ICH M7 is proposing to predict potential DNA reactivity and thus mutagenicity of all impurities where the structure is known by using in silico (quantitative) structure-activity relationship (Q) SAR methodologies.

Impurities that do not trigger any structural mutagenicity alert in an appropriate (Q)SAR assessment would be considered as non-mutagenic. When a structural alert for mutagenicity is identified the impurity can either be controlled as a mutagenic impurity or an Ames test can be conducted in order to verify the in silico prediction. A negative ICH-compliant Ames test with the impurity alone would overrule any structure-based concern and the impurity would be considered non-mutagenic. For impurities that are not feasible to isolate or synthesize or when compound quantity is limited, it may not be possible to achieve the highest test concentrations recommended for an ICH-compliant test. In this case, bacterial mutagenicity testing could be carried out using a miniaturized assay format with proven high concordance to the ICH-compliant assay to enable testing at higher concentrations with justifications.

4.3.5 Structure-Based Assessment of Impurities: QSAR Methodology

Mutations are generally caused by an interaction of reactive chemicals with the DNA bases. A wealth of knowledge has been generated to date to characterize the chemistry around such reactions. A limited number of functional groups have been identified to efficiently confer such reactions and Ashby and Tennant [20] have been the leading experts to publish such evidence more than 25 year ago. To date, their basic list of dangerous functional groups has seen only minor changes. Some of these functional groups are generated via metabolic conversions in the body and may be very short-lived. Very clearly, the reactive chemistry involved in the process of formation of adducts at DNA bases often is similar to the chemistry used to efficiently synthetize small molecule APIs.

Based on this knowledge, structure-based assessments are nowadays commonly performed with the support of in silico (Q)SAR models/systems on all compounds used and formed during the synthesis process (i.e., starting materials, reagents, synthesis intermediates) that may be present in the drug substance, and likely or plausible by-products and degradants. Structural Alerts (SA) for Ames mutagenicity are applied to detect DNA reactivity of impurities. Since clastogenicity alerts cover effects at the chromosomal level that may occur through indirect mechanisms likely to have a threshold (see Sect. 4.3.1), this class of SA does not play a role in the characterization of potentially mutagenic impurities. Also, carcinogenicity alerts that might also identify non-genotoxic carcinogens are generally not used for the evaluation of mutagenic impurities, considering that the scope is restricted to DNA-reactive and potential mutagenic carcinogens.

In silico methods to support the prediction of mutagenic activity have been available for more than 20 years and have been continuously improved. In a recent paper [21], their use in the pharmaceutical industry has been reviewed with the aim to (1) clarify the place/use of (Q)SAR models in the structure-based assessment of potentially mutagenic impurities, (2) highlight the quality criteria for (Q)SAR models to be used for the evaluation of impurities and possibly reach a consensus on recommendations for users, and (3) enhance the transparency of the process. The most commonly used in silico tools can be divided into two categories: (1) empirical or rule-based expert knowledge systems; (2) statistical or Quantitative Structure Activity Relationship (QSAR)-based techniques. The commonly used tools (DEREK, MultiCase/MC4PC, SciQSAR, Leadscope Model Applier, OECD toolbox, ToxTree) were summarized by Sutter et al. [21]. Overall, it appears that the current approaches for conducting structure-based assessments successfully predict compounds as being non-mutagenic, i.e. Ames negatives (negative predictive value of at least 86 %, sensitivity of 80–95 %). Given that the reproducibility of the Ames assay has been reported as only 87 % [22], a further increase in sensitivity or negative predictive value of structure-based approaches would be disproportionate and is therefore not needed. Most pharmaceutical companies tend to use a rule-based expert system such as DEREK as their primary in silico tool. Additional (Q)SAR systems, either commercially available ones or in-house developments are usually used to complement this basic prediction layer. This is in line with recommendations made in the ICH M7 guideline, which stipulates the use of two systems using different methodologies to corroborate each other and to arrive at a level of confidence that predictions comply with the expectations of Health Authorities. This is important as any prediction leading to a "no alert" call will generally imply the end of the search process at this stage with no further need for generation of experimental data.

4.3.6 Classification of Impurities with Respect to Mutagenic and Carcinogenic Potential

The appropriate way for controlling a (potentially) mutagenic impurity depends very much on the extent of toxicological data available for the chemical under review. To this end ICH M7 has adapted a classification system proposed by Müller et al. [9] defining five separate classes of impurities with respect to data availability for defining their mutagenic and carcinogenic potential.

Class 1 impurities include known (rodent) carcinogens with a likely mutagenic mode of carcinogenic action as usually indicated by positive results from a bacterial mutagenicity assay. For impurities in this class a compound-specific risk assessment is recommended to derive acceptable intakes (see Sect. 4.3.7.1).

Class 2 impurities include experimentally established mutagens, i.e., usually Ames positives with no data from standard carcinogenicity studies available and therefore with unknown carcinogenic potential. For impurities of this class an acceptable intake level for lifetime daily exposure would be based on the generic TTC.

Class 3 impurities include compounds with structural alerts for mutagenicity that are unrelated to the structure of the drug substance and for which no mutagenicity data are available. Such impurities would be considered as mutagenic and would thus be controlled like Class 2 impurities. If tested for bacterial mutagenicity Class 3 impurities would switch either into Class 2 in case of a positive result or into Class 5 when testing is negative.

Class 4 impurities contain a structural alert for mutagenicity which, however, is shared by the drug substance or compounds related to the drug substances such as a process intermediate which have been tested negative for mutagenicity.

Class 5 impurities include compounds with no structural alerts or where structural alerts have been identified but these alerts were overruled by sufficient data to demonstrate lack of mutagenicity or carcinogenicity.

Class 4 and 5 impurities would be controlled like ordinary impurities according to ICH Q3A/B.

4.3.7 Compound-Specific Risk Assessment: Adjustment of Acceptable Intake Levels

The TTC concept as described above is intended to be applied to (potentially) mutagenic impurities that lack carcinogenicity data. If any such data are available, they should be used for estimating a compound-specific acceptable daily intake value corresponding to a 10^{-5} lifetime risk of cancer. The range of carcinogenic potencies spans at least six orders of magnitude; for instance, the TD50s in the CPDB of mitomycin and phenacetin are 0.001 and 1250 mg/kg/day, respectively. Since the TTC of 1.5 μg/day has been established to be protective for the vast majority of the CPDB carcinogens it is clear that applying the generic TTC value would be excessively conservative for most of them. On the other hand, a few rodent carcinogens (besides those belonging to the Cohort of Concern) are of such high potency that intake levels below 1.5 μg/day would be required.

In addition to the use of carcinogenicity data of the impurity itself to perform a compound-specific assessment it may also be possible to apply a read-across approach and use carcinogenicity potency data of close analogues of the impurity under investigation and estimate a class-specific potency. This class-specific carcinogenic potency value may then be used to adjust acceptable intake levels of untested chemicals belonging to the same class. Also, availability of in vivo mutagenicity data may justify deviation from generic TTC for an Ames positive impurity without carcinogenicity data. Case examples of these different data scenarios and their use to justify a compound-specific acceptable intake level will be discussed in the next sections.

4.3.7.1 Adjustment Based on Compound-Specific Carcinogenicity Data

For Ames-positive impurities with rodent carcinogenicity data (Class 1 impurities) the ICH M7 guideline recommends in Note 4 the calculation of a compound-specific acceptable intake (AI) from the TD_{50} using the following formula:

$$AI = TD_{50}\left[mg / kg\,b.w. / day\right] / 50{,}000 \times 50\left[kg\,b.w.\right]$$

This calculation represents a similar linear extrapolation as has been used for deriving the generic TTC (see Sect. 4.3.3). The weight adjustment assumes an arbitrary adult human body weight for either sex of 50 kg. This relatively low weight provides an additional safety factor against the standard weights of 60 or 70 kg that are often used in this type of calculation.

To illustrate this approach the rodent carcinogenicity data of ethylene oxide are used as an example in Note 4 of the ICH M7 guideline. The CPDB displays two TD50 values for ethylene oxide; 21.3 mg/kg/day from a rat study and 63.7 mg/kg/day from a mouse study. The calculation of an acceptable intake is done with the lower and therefore more conservative value using the above formula and results in an acceptable daily life-long intake level of 21.3 μg ethylene oxide. For a daily therapeutic dose of a drug substance of, for instance, 100 mg this specific AI results in a maximum allowable concentration of 210 ppm ethylene oxide as an impurity in the drug substance. The ICH M7-derived limit is considerably higher than the maximum limit of 1 ppm as requested by the EU Note for Guidance [4] as discussed earlier in Sect. 4.1. This example demonstrates very well the paradigm shift in regulatory thinking that has occurred over the last decade towards a toxicologically defined acceptance of low levels of DNA reactive carcinogens in lieu of a control policy based on ALARP principle and analytical capabilities.

As an alternative to the default approach of using the most conservative TD50 value (most sensitive species, most sensitive organ) ICH M7 is proposing an in-depth toxicological expert assessment of the available carcinogenicity data in order to initially identify the most important findings (target species, tumor site, mode of action etc) of human relevance as a basis for deriving acceptable intake values.

Cases falling into this category would be Ames positive rodent carcinogens with sufficient evidence for a threshold mode of action for the tumors. Calculation of a permissible daily exposure (PDE) rather than linear extrapolation of an AI would be considered appropriate as is illustrated by the following example epichlorohydrin.

Epichlorohydrin is a strong alkylating agent used as an intermediate in the manufacturing of pharmaceuticals. It is a direct-acting mutagen with positive results in the Ames test [23, 24] and known to produce forestomach tumors in rats following drinking water exposures [25, 26]. However, epichlorohydrin also causes nasal cavity tumors following inhalation exposure [27] and as such is described as a 'site-of-contact' carcinogen. The most likely mode of action for induction of forestomach tumors only (no tumors in other tissues) following oral administration of highly irritating compounds is a direct cytotoxic effect at the site of first contact leading to irritation or ulceration and ultimately occurrence of tumor when exposure is chronic [28]. The highly irritating potential of epichlorohydrin has been shown in short-term oral rat studies where a dose-related induction of lesions of the forestomach was observed [29]. It is therefore highly unlikely that epichlorohydrin poses a risk of tumors in humans exposed to low concentrations as impurities in pharmaceuticals, well below concentrations that could cause irritation or inflammation. For carcinogens with a threshold mode of action it is justified to use the PDE calculation according to the methodology explained in the ICH Q3C guideline. The oral 2-year rat study [26] can be used to calculate the PDE. Since a NOEL is not established, the LOAEL of 2 mg/kg/day for rat forestomach tumors and hyperplasia is used with an uncertainty factor F5 of 10 (ICH Q3C guideline, Appendix 3).

$$\text{PDE} = (\text{LOAEL}) \times 50 \text{ kg body wt.} / (\text{F1} \times \text{F2} \times \text{F3} \times \text{F4} \times \text{F5})$$

F1 = 5 (rat to human);
F2 = 10 (individual variability);
F3 = 1 (more than half lifetime);
F4 = 10 (severe toxicity; tumors);
F5 = 10 (NOEL not established)

$$\begin{aligned} \text{PDE} &= (2 \text{ mg/kg/day}) \times (50 \text{ kg}) / (5 \times 10 \times 1 \times 10 \times 10) \\ &= 100/5000 \text{ mg/day} = 20 \mu\text{g/day}. \end{aligned}$$

The lifetime PDE of epichlorohydrin would thus be 20 μg.

4.3.7.2 Adjustment Based on Class-Specific Carcinogenicity Data (Read-Across Approach)

Knowledge of the carcinogenic potency of certain structural classes may allow use of more knowledge-based calculations of acceptable intake values, in preference to the default TTC value. For instance, this approach has been used to identify the Cohort-of-Concern as a class of carcinogens that were considered extremely potent and may still be of concern even at intakes below the TTC [18]. According to ICH M7 the acceptable intakes for these high-potency carcinogens would "likely be significantly lower" than the default TTC.

On the other hand the ICH M7 guideline is proposing an approach in which a chemically-defined class or sub-class of known carcinogens with low(er) carcinogenic potency is defined that may allow calculation of safe intake levels that is higher than the default TTC.

Monofunctional alkyl chlorides, a group of chemicals commonly used in drug syntheses, is presented as an example in Note 5 of ICH M7. Brigo and Müller [30] identified 27 alkyl chlorides with a molecular weight of less than 250 that had mutagenicity data and calculated carcinogenic potencies (TD50). The monofunctional compounds (with no other alkylating function) were found to be much less potent than those in the multifunctional sub-class. The 14 monofunctional alkyl chlorides had potencies that translate into acceptable intakes of 36–1810 μg/day (TD50 [mg / kg b.w. / day] / 50,000 × 50 [kg b.w.]). Thus, the calculated acceptable daily intakes corresponding to the same negligible risk level as the default TTC are at least 20 times higher than the default lifetime TTC of 1.5 μg. ICH M7 is proposing a more conservative sub-class-specific lifetime limit, ten times higher than the defaults that can be used for any mutagenic monofunctional alky chloride for which no carcinogenicity data are available.

The read-across approach from cancer potencies of a defined category of known carcinogens to a single untested mutagenic impurity may also be applied to define an acceptable limit for a compound of the Cohort of Concern.

Example N-Nitrosomorpholine Compounds
N-Nitrosothiomorpholinedioxide has been determined as an impurity in a drug substance. N-Nitrosothiomorpholinedioxide is likely carcinogenic with a genotoxic mode of action based on its structural analogy to other N-nitroso-compounds and, specifically, to nitrosomorpholines (see below). N-nitroso-compounds are not directly genotoxic but require metabolic activation to generate the ultimate electrophile. Since N-nitroso-compounds belong to the cohort of concern the default TTC of 1.5 μg/day is not applicable to control N-nitrosothiomorpholinedioxide. ICH M7 is proposing in this situation "a case-by-case approach using e.g., carcinogenicity data from closely related structures, to justify acceptable intakes." (ICH M7, Sect. 7.5). The CPDB contains TD50 values for a number of close structural analogues of N-nitrosothiomorpholinedioxide which can be used to roughly estimate the cancer potency specific for this structurally defined sub-class (Table 4.1).

There appears to be regulatory precedence to base risk assessment in such cases on the worst structural analogue in terms of carcinogenic potency. In our case, the most potent carcinogenic analogue of N-nitrosothiomorpholinedioxide is nitrosomorpholine for which an AI of 0.11 μg/day has been calculated which is roughly 15-fold lower than the generic TTC level of 1.5 μg.

4.3.7.3 Adjustment Based on Compound-Specific In Vivo Mutagenicity Data

Ames-positive impurities without compound-specific carcinogenicity data (Class 2) may have vivo mutagenicity data and the interesting question is whether and, if so, how such data could be used for a compound-specific assessment thus deviating from the Class 2 control approach.

There are two different scenarios on how in vivo mutagenicity data may be used in this context: (i) negative in vivo mutagenicity findings may provide sufficient evidence for concluding that the vitro mutagenicity findings in bacteria are non-relevant under in vivo conditions and would support setting compound-specific impurity limits higher than the default TTC; (ii) the in vivo mutagenicity studies provide convincing evidence for a threshold mechanism of mutagenesis and would support a PDE calculation rather than linear extrapolation from TD50 to define a safe impurity limit.

Table 4.1 TD 50 values and related Acceptable Intakes (AIs) of close structural analogues to nitrosothiomorpholinedioxide

Analogous structures	TD50 in rat (mg/kg b.w./day)	Lifetime AIs TD50 (mg/kg b.w./day) /50,000 × 50 (kg b.w.)
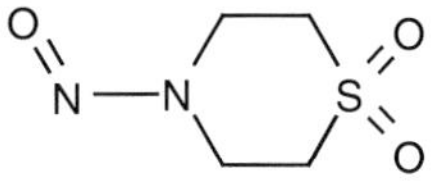 Nitrosothiomorpholine-dioxide	No carcinogenicity data (impurity under review)	?? Lowest sub-class-specific AI may be used (see text)
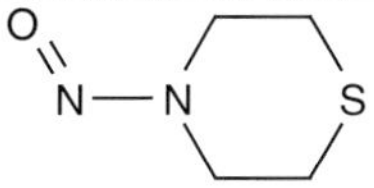 Nitrosothiomorpholine	5.39 mg/kg	5.4 μg
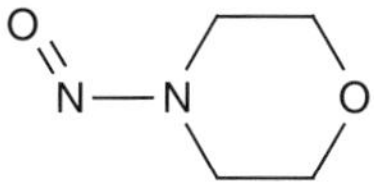 Nitrosomorpholine	0.109 mg/kg	0.11 μg
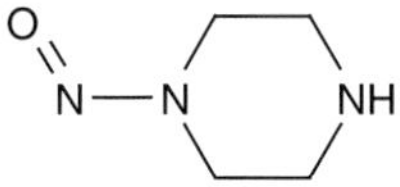 Nitrosopiperazine	8.78 mg/kg	8.8 μg
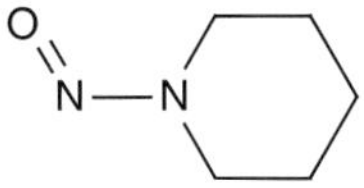 Nitrosopiperidine	1.43 mg/kg	1.4 μg
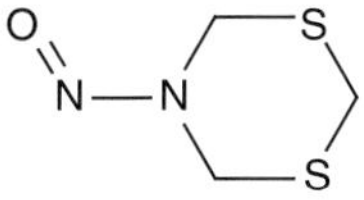 Nitrosodithiazinane	Not carcinogenic	–

A de-risking strategy of using in vivo mutagenicity studies is proposed in ICH M7 for such rare cases when due to technical limitations an Ames-positive impurity cannot be controlled at limits recommended for Class 2 impurities. In particular Note 3 gives guidance, which in vivo follow up tests are potentially suitable to better characterize the mutagenic potential of the Ames positive impurity under question. While the transgenic gene mutation assay is considered as "golden standard" to follow up an Ames positive, other in vivo models such as the Pig-a assay, the micronucleus test, the rat liver UDS test or the comet assay can be acceptable if the choice of test as fit-for-purpose is sufficiently justified.

This path can also be advisable if the impurity structure belongs to a certain structural group, which is known to have limited predictive value to exert a mutagenic potential in vivo or to possess carcinogenic potential. Aromatic amines, for which complex metabolism pathways are to be expected and tissue-specific tumorigenic effects are well known, belong to such a class. For this group of intermediates/impurities, often an additional consideration comes into play in that they might emerge as a metabolite of the API. For instance, it is common practice to pursue linkage of ring systems for pharmaceutical candidate compounds with an amide bond, which involves aromatic amine intermediates. While the aromatic amine involved for generation of this bond represents a risk for contamination of the API as an impurity, the cytosolic hydrolases contained in our cells cleave this bond forming the aromatic amine as a metabolite. Normally, the amount of the aromatic amine generated by metabolism pathways clearly outweighs the level of contamination of the API with the same entity as an impurity. While impurity levels can usually be controlled, the risk stemming from metabolism is unavoidable. In this situation, follow up testing to better characterize the mutagenic/carcinogenic risk of such an aromatic warrants spending of sometimes substantial resources.

The policy of some companies includes the avoidance of such situations via upfront testing of the aromatic amine components to select non-mutagenic ones. However, in case appropriate non-mutagenic structures are not available, a strategy to follow-up the Ames positive result may include an in vivo transgenic mutation study or the Pig-a assay. The Pig-a assay is undergoing interlaboratory assessment for validity and warrants regulatory acceptance upon availability of a standard protocol. Negative results in these models would support further development of the drug candidate. A consideration in such cases is whether to test the aromatic amine directly or to test the drug candidate up to the MTD and rely on the proficiency of the test organism (mouse or rat) to generate this metabolite. The latter approach generates a better simulation of the metabolic pathway of concern and may even lead to a higher target tissue exposure to the aromatic amine than if it would be administered orally as defined entity. An additional advantage of testing the API is that one avoids an in vivo exposure to potentially problematic levels of impurities in the aromatic amine impurity, which would have to be available in considerable amounts, if tested on its own.

In the context of an accident, which occurred in the production of the HIV antiviral nelfinavir in 2007, an extensive set of in vivo mutagenicity studies was conducted for the alkylating agent ethyl methanesulfonate (EMS). The results of these studies were supportive of the assumption of a non-linear dose–response with several doses being non-discernable from the background mutation level [31, 32]. Thus, they supported regulatory evaluations and risk management decisions based on the no-observed-effect-levels from these studies. In later studies, the observations were generally confirmed, albeit in some models on a lower dose level [33]. These later studies confirm the need for remaining conservative on the regulatory side with risk assessment and risk management procedures in such cases.

4.3.8 Less-Than-Lifetime Approaches

As described, Müller et al. [9], were the first to describe an adjustment of the lifetime risk TTC model to the needs of shorter durations of intake/exposure commonly encountered in clinical trials for pharmaceuticals. Also, the approved conditions under which a specific pharmaceutical is taken to combat disease may often not be lifetime.

The concept of accepting higher intake levels for mutagenic impurities in pharmaceuticals for indications with less-than-lifetime (LTL) treatment duration is based on the assumption that cancer risk increases as a function of cumulative dose. This allows higher daily intake of mutagenic impurities for LTL durations as long as the cumulative dose does not exceed the acceptable cumulative lifetime dose. This concept is consistent with Haber's Rule developed in the 1920s where the toxic outcome (k) was related to concentration (or dose) of the toxic chemical and time of exposure or $C \times T = k$. Haber based this rule on his inhalation studies with poisonous gases used as weapons in World War I [34] and his observation that the product of exposure concentration and time resulted in a constant lethal response. This basic concept is nowadays also applied in cancer risk assessment for genotoxic carcinogens. Based on the stochastic nature of the carcinogenic process the probability of a tumor is considered to be proportional to the total number of molecules at the target site and thus proportional to the total or cumulative dose. However, experimental evidence for this correlation is scarce. Only a limited number of animal studies have assessed the comparative tumor incidence from short-term versus long-term exposures with similar cumulative doses (reviewed in [35, 36]). In general such studies suggest that linear extrapolation from lifetime exposure experiments to a short-term exposure can lead to both an underestimation and overestimation of the cancer risk, but that an underestimation is more likely. It can be anticipated that the more the lifetime cumulative dose is compressed into a shorter time period and thus resulting in a higher dose rate, the greater the possibility that risk will be underestimated. In order to adjust for such uncertainties it has been proposed to add a correction factor to the $C \times T$ rule when calculating acceptable short-term exposures to mutagenic carcinogens. For instance, Verhagen et al. [37] suggest that the risk associated with a peak exposure can be estimated by using the total lifetime dose and applying a correction factor of 10. It is not known, however, how the dose-responses would look at very low cumulative exposures, an exposure to which the risk assessment for impurities is projected to. Similarly, the experiments, in which the cumulative exposure was compressed into a shorter than lifetime treatment duration, have been done in young animals. Hence, the impression of an increased risk for tumors under such conditions, compared to a lifetime experiment is likely confounded by the higher risk of mutation-related tumor induction early in life than later in life.

Nevertheless, a similar correction factor approach has been applied in the ICH M7 guideline for setting LTL-limits of mutagenic impurities. Based on the accepted lifetime TTC of 1.5 μg/day and a lifetime of 70 years (=25,550 days) the calculated acceptable cumulative lifetime dose is: 1.5 μg/day × 25,550 days = 38,250 μg. Hence, the calculated intake level of a single dose (per lifetime) corresponding to a lifetime cancer risk of 10^{-5} would theoretically be 38.3 mg. Similarly, the calculated daily intake levels would be 10 μg for 10 years, 100 μg for 1 year and 1270 μg for 1 month, all resulting in the same cumulative intake of 38.3 mg and therefore theoretically in the same cancer risk. The acceptable LTL intakes that are in fact recommended in ICH M7 are 10 μg (>1–10 years), 20 μg (>1–12 months) and

120 µg (up to 1 month) and thus take into account correction factors which are higher for shorter treatment durations. For instance, the acceptable single dose of a mutagenic impurity of 120 µg as recommended in ICH M7 is about 300-fold lower than calculated CxT value without correction factor. These proposed LTL intakes can be applied to both, investigational drugs during clinical development as well as marketed products when limited treatment durations can be anticipated. The acceptable intakes derived from compound-specific risk assessment as described in Sect. 4.3.7 can also be adjusted for shorter duration of use in the same proportions as described before.

4.4 Future Impact

One of the main impacts of the standards set forth by the ICH M7 guideline is a seemingly broad application of the TTC concept to an area, for which it was originally not foreseen, i.e. carcinogenesis as induced by agents causing mutations. Further, the approach to allow for higher intake of such agents for a shorter period than lifetime was also not applied anywhere for regulatory purposes so far. It will be interesting to see how this concept may be used elsewhere than for impurities in pharmaceuticals, e.g. for residues in food, ingredients in biologicals, multiple intake situations, plant-derived pharmaceuticals, etc. Also, progress is being made on the clarification of no-biological-effect levels for mutagenic carcinogens [31, 32], which will probably add to an improved projection of risk assessment into low dose exposure situations. As the pharmaceutical industry has been confronted for more than 10 years with the TTC-based risk assessment and control approach for mutagenic impurities it is predicted that ICH M7 is being smoothly integrated into the pharmaceutical risk assessment and quality control business.

References

1. ICH (2006a) Harmonised Tripartite Guideline Q3A(R2): impurities in new drug substances. http://www.ich.org/fileadmin/Public_Web_Site/ICH_Products/Guidelines/Quality/Q3A_R2/Step4/Q3A_R2__Guideline.pdf. Accessed 26 Nov 2014
2. ICH (2006b) Harmonised Tripartite Guideline Q3B(R2): impurities in new drug products. http://www.ich.org/fileadmin/Public_Web_Site/ICH_Products/Guidelines/Quality/Q3B_R2/Step4/Q3B_R2__Guideline.pdf. Accessed 26 Nov 2014
3. ICH (2011) Harmonised Tripartite Guideline Q3C(R5): impurities: guideline for residual solvents. http://www.ich.org/fileadmin/Public_Web_Site/ICH_Products/Guidelines/Quality/Q3C/Step4/Q3C_R5_Step4.pdf. Accessed 26 Nov 2014
4. EMEA (2001) European Agency for the Evaluation of Medicinal Products. Note for Guidance on limitations to the use of ethylene oxide in the manufacture of medicinal products. http://www.ema.europa.eu/docs/en_GB/document_library/Scientific_guideline/2009/09/WC500002915.pdf. Accessed 14 Nov 2014

5. IARC International Agency on the Research of Cancer (2012) Ethylene oxide. IARC Monogr Eval Carcinog Risks Hum 100F:379–400
6. EMA (2006) European Medicines Agency. Guideline on the limits of genotoxic impurities. http://www.ema.europa.eu/docs/en_GB/document_library/Scientific_guideline/2009/09/WC500002903.pdf. Accessed 26 Nov 2014
7. EMEA (2002) European Agency for the Evaluation of Medicinal Products. Draft CPMP Position Paper on the limits of Genotoxic Impurities (CPMP/SWP/5199/02/draft2, 18 December 2002)
8. EMA (2010) European Medicines Agency. Question and answer on the 'Guideline on the limits of genotoxic impurities'. EMEA/CHMP/SWP/431994/2007 Rev. 3. http://www.ema.europa.eu/docs/en_GB/document_library/Scientific_guideline/2009/09/WC500002907.pdf. Accessed 26 Nov 2014
9. Müller L, Mauthe RJ, Riley CM et al (2006) A rationale for determining, testing, and controlling specific impurities in pharmaceuticals that possess potential for genotoxicity. Regul Toxicol Pharmacol 44:198–211
10. U.S. Food and Drug Administration, FDA (2008) Draft guidance for industry on genotoxic and carcinogenic impurities in drug substances and products: recommended approaches. http://www.fda.gov/ohrms/dockets/98fr/fda-2008-d-0629-gdl.pdf (Accessed 09 Sept 2015)
11. ICH (2014) Harmonised Tripartite Guideline M7: assessment and control of DNA reactive (mutagenic) impurities in pharmaceuticals to limit potential carcinogenic risk. http://www.ich.org/fileadmin/Public_Web_Site/ICH_Products/Guidelines/Multidisciplinary/M7/M7_Step_4.pdf. Accessed 26 Nov 2014
12. Kirkland D, Aardema M, Henderson L et al (2005) Evaluation of the ability of a battery of 3 *in vitro* genotoxicity tests to discriminate rodent carcinogens and non-carcinogens. I. Sensitivity, specificity and relative predictivity. Mutat Res 584:1–256
13. Rulis AM (1989) Establishing a threshold of regulation. In: Bonin JJ, Stevenson DE (eds) Risk assessment in setting national priorities. Plenum, New York, pp 271–278
14. Food US, Drug Administration FDA (1995) Food additives: threshold of regulation for substances used in food-contact articles (final rule). Fed Regist 60:36582–36596
15. Gold LS, Sawyer CB, Magaw R et al (1984) A carcinogenic potency database of the standardized results of animal bioassays. Environ Health Perspect 58:9–319
16. Cheeseman MA, Machuga EJ, Bailey AB (1999) A tiered approach to threshold of regulation. Food Chem Toxicol 37:387–412
17. Munro IC (1990) Safety assessment procedures for indirect food additives: an overview. Report of a workshop. Regul Toxicol Pharmacol 12:2–12
18. Kroes R, Renwick AG, Cheeseman M et al (2004) Structure-based threshold of toxicological concern (TTC): guidance for application to substances present at low levels in the diet. Food Chem Toxicol 42:65–83
19. Kenyon MO, Cheung JR, Dobo KL et al (2007) An evaluation of the sensitivity of the Ames assay to discern low-level mutagenic impurities. Regul Toxicol Pharmacol 48:75–86
20. Ashby J, Tennant RW (1988) Definitive relationships among chemical structure, carcinogenicity and mutagenicity for 301 chemicals tested by the US NTP. Mutat Res 257:229–306
21. Sutter A, Amberg A, Boyer S et al (2013) Use of in silico systems and expert knowledge for structure-based assessment of potentially mutagenic impurities. Regul Toxicol Pharmacol 67:39–52
22. Kamber M, Flückiger-Isler S, Engelhardt G et al (2009) Comparison of the Ames II and traditional Ames test responses with respect to mutagenicity, strain specificities, need for metabolism and correlation with rodent carcinogenicity. Mutagenesis 24:359–366
23. Canter DA, Zeiger E, Haworth S et al (1986) Comparative mutagenicity of aliphatic epoxides in *Salmonella*. Mutat Res 172:105–138
24. Giri AK (1997) Genetic toxicology of epichlorohydrin: a review. Mutat Res 386:25–38
25. Konishi Y, Kawabata A, Denda A et al (1980) Forestomach tumors induced by orally administered epichlorohydrin in male Wistar rats. Gann 71(6):922–923

26. Wester PW, Van der Heijden CA, Bisschop A et al (1985) Carcinogenicity study with epichlorohydrin (CEP) by gavage in rats. Toxicology 36:325–339
27. Laskin S, Sellakumar AR, Kuschner M et al (1980) Inhalation carcinogenicity study of Epichlorhydrin in noninbred Sprague–Dawley rats. J Natl Cancer Inst 65:751–757
28. Proctor DM, Gatto NM, Hong SJ et al (2007) Mode-of-action framework for evaluation the relevance of rodent forestomach tumors in cancer risk assessment. Toxicol Sci 98:313–326
29. Daniel FB, Robinson M, Olson GR et al (1996) Toxicity studies of epichlorohydrin in Sprague-Dawley rats. Drug Chem Toxicol 19:41–58
30. Brigo A, Müller L (2010) Development of the threshold of toxicological concern concept and its relationship to duration of exposure. In: Teasdale A (ed) Genotoxic impurities: strategies for identification and control. Wiley, New York, pp 27–63
31. MacGregor J, Frötschl R, White PA et al (2014) IWGT report on quantitative approaches to genotoxicity risk assessment II. Use of Point-of-Departure (PoD) metrics in defining acceptable exposure limits and assessing human risk. Mutat Res. doi:10.1016/j.mrgentox.2014.10.008
32. Müller L, Gocke E, Lave T et al (2009) Ethyl methanesulfonate toxicity in Viracept – a comprehensive human risk assessment based on threshold data for genotoxicity. Toxicol Lett 190:317–329
33. Cao X, Mittelstaedt RA, Pearce MG, Allen BC, Soeteman-Hernandez LG, Johnson GE, Bigger CA, Heflich RH (2014) Quantitative dose–response analysis of ethyl methanesulfonate genotoxicity in adult gpt-delta transgenic mice. Environ Mol Mutagen 55:385–399
34. Haber F (1924) Zur Geschichte des Gaskrieges. In: Haber F (ed) Fünf Vorträge aus den Jahren 1920–1923. Springer, Berlin, pp 23–29
35. Bos PMJ, Baars B-J, van Raaij MTM (2004) Risk assessment of peak exposure to genotoxic carcinogens: a pragmatic approach. Toxicol Lett 151:43–50
36. Felter SP, Conolly RB, Bercu JP et al (2011) A proposed framework for assessing risk from less-than-lifetime exposures to carcinogens. Crit Rev Toxicol 41:507–544
37. Verhagen H, Feron VJ, van Vliet PW (1994) Risk assessment of peak exposure to genotoxic carcinogens. Health Council of the Netherlands, Hague, Publication no. A94/04

Chapter 5
New and Emerging Genetic Toxicity Tests and Approaches to Genetic Toxicology Testing

Patricia A. Escobar, Stephen D. Dertinger, and Robert H. Heflich

Abstract Genetic toxicology tests are conducted to determine if a compound (drug) has the potential to cause mutations or chromosomal damage, data that has value in predicting its carcinogenic potential. For new pharmaceuticals, test data from a standard test battery are required prior to administering the compound to humans. This test battery includes in vivo studies of chromosomal damage in the rodent hematopoietic system, traditionally performed in the bone marrow, but in recent years, the in vivo micronucleus assay has been validated for use in peripheral blood. In addition, new regulatory guidelines, like ICH S2R1, allow the use of a second in vivo assay as part of the genotoxicity assessment, and having different in vivo assays to use will benefit understanding the in vivo genotoxicity profile of candidate pharmaceuticals. Three 'new' in vivo tests are described—the in vivo Comet assay, the in vivo *Pig-a* gene mutation assay, and the liver micronucleus (MN) assay. An established test, the transgenic rodent (TGR) gene mutation assay, is presented as a test whose use for regulatory decision-making may increase due to the recent introduction of robust testing protocols and new test guidelines. The chapter also discusses refinements in how the traditional genotoxicity tests are conducted and interpreted, including new ways of following up on their findings. Finally, the chapter discusses what has been described as a paradigm shift—that is, a move away from qualitative categorizations of genotoxic vs. non-genotoxic to the use of quantitative genotoxicity data as an adverse outcome on which regulatory decisions may be made.

Disclaimers and Disclosures The information in these materials is not a formal dissemination of information by the U.S. Food and Drug Administration and does not represent agency position or policy. S.D.D. is an employee of Litron Laboratories; Litron holds patents, sells commercial kits, and performs fee-for-service work in the area of flow cytometry-based genotoxicity assessments.

P.A. Escobar
Merck Research Laboratories, West Point, PA, USA
e-mail: patricia.escobar1@merck.com

S.D. Dertinger
Litron Laboratories, Rochester, NY 14623, USA

R.H. Heflich (✉)
Division of Genetic and Molecular Toxicology, National Center for Toxicological Research/U.S. Food and Drug Administration (U.S. FDA/NCTR), 3900 NCTR Rd., Jefferson, AR 72079, USA
e-mail: robert.heflich@fda.hhs.gov

M.J. Graziano, D. Jacobson-Kram (eds.), *Genotoxicity and Carcinogenicity Testing of Pharmaceuticals*, DOI 10.1007/978-3-319-22084-0_5

Keywords Comet assay • *Pig-a* assay • Liver micronucleus • Transgenic rodent • Follow-up testing

5.1 Introduction

Mutation was recognized as a potential human health hazard beginning at least in the middle of the twentieth century (reviewed in [1,2]). However, it was reports in the 1970s indicating that carcinogens are mutagens and that relatively simple short-term *in vitro* mutagenicity tests in bacteria can be used to identify suspect carcinogens [3,4] that provided the major stimulus that resulted in use of genetic toxicology data for regulatory safety assessments. Further scientific support for the use of genetic toxicology tests as predictors of carcinogenic potential came from subsequent discoveries on the roles of oncogenes and mutated tumor suppressor genes in cancer, thus providing a mechanistic link between the mutagenicity of chemicals and their ability to cause cancer. Regulatory agencies quickly adopted genotoxicity tests as aids in fulfilling their regulatory responsibilities to limit human exposure to carcinogens, providing recommendations for their use [5].

Despite the growing scientific acceptance and widespread application of *in vitro* mutagenicity tests as screens for potential carcinogenicity, it was realized early on that no one test is sufficient for detecting all the various types of mutations and DNA damage responsible for mutation, and that the simplest tests, *i.e.,* those conducted in bacteria, may not be sufficient for measuring the risks to intact animals. Thus, tests were grouped together into genotoxicity testing batteries to better evaluate the potential hazards associated with test substances [2,6]. The battery used for pharmaceutical testing was formalized in 1997 in the International Conference on Harmonization of Technical Requirements for Registration of Pharmaceuticals for Human Use (ICH) S2B test guidance [7]. This battery (since revised) included a bacterial test for evaluating gene mutation, an *in vitro* mammalian cell test for chromosomal damage or the *in vitro* mouse lymphoma gene mutation assay (which is also sensitive to chromosome damage), and an in vivo test for chromosome damage using rodent hematopoietic cells.

As results from the new genetic toxicology tests began to accumulate, it became clear that the predictive power of individual tests and of test batteries was not perfect, at least for cancer induction in rodent bioassays, the usual 'gold standard'. Studies by the National Toxicology Program in the 1980s revealed that there are a relatively large number of rodent carcinogens that are not mutagens [8]. These non-genotoxic carcinogens induce tumors by alternative pathways (*e.g.*, α2 μ-globulin nephropathy, peroxisome proliferation, hormonal effects, inflammation, calculus formation, etc.) that may or may not be relevant for humans. It also has become clear that adding more *in vitro* tests to the genotoxicity test batteries may increase the sensitivity for detecting DNA damaging agents at the expense of specificity for identifying carcinogens [9]. Added to this is a growing body of evidence that the relationship between mutations and cancer is complex, with carcinogenesis often involving multiple clones of cells containing different cancer-related mutations, some of which are present at high frequency in healthy tissue [10]. While all of these factors complicate the simple relationship between mutation and cancer, industry and regulatory scientists still rely on genetic toxicology data for making decisions on product development and human safety.

In recent years there has been a gradual refinement in the way genotoxicity testing is conducted in order to provide better predictions of human health risk. For pharmaceutical testing, there is a trend towards more *in vivo* testing and for integrating *in vivo* genotoxicity endpoints into general toxicology studies, as reflected in the recently revised ICH S2 (R1) guidance [11]. Ideally this approach can generate a robust set of data that combines multiple *in vivo* genotoxicity endpoints into the same study, potentially integrating the assays into on-going toxicology studies [12] in a manner that is consistent with the 3Rs principles.

In this chapter we describe three ‘new’ *in vivo* tests, the *in vivo* Comet assay, the *in vivo Pig-a* gene mutation assay and the liver micronucleus (MN) assay, that are at various stages of development and regulatory acceptance. In addition, an older genetic toxicology test, the transgenic rodent (TGR) gene mutation assay, is presented as a test whose use for regulatory decision-making may increase due to the recent introduction of robust testing protocols and new test guidelines. These tests provide additional tools for assessing genetic toxicity *in vivo* beyond the classical *in vivo* cytogenetic assays by generating data using additional tissues, a requirement of ICH S2 (R1) Option 2, and using additional genotoxicity endpoints. This chapter also will discuss refinements in how the traditional genotoxicity tests are conducted, and interpreted, including new ways of following up on their findings. Finally, the chapter will discuss the movement towards rethinking the paradigm under which genetic toxicity testing is conducted by using genetic toxicology data in a quantitative manner and shifting the emphasis on using genotoxicity data only for identifying carcinogens to considering genotoxicity *per se* as an adverse outcome on which regulatory decisions are made.

5.2 Comet Assay

The Comet assay, also known as single cell gel elcctrophoresis (SCGE), is a microgel electrophoresis technique that detects DNA damage, by measuring DNA strand breaks in individual cells. This technique was first introduced by Ostling and Johanson [13], and involved embedding individual mammalian cells in an agarose gel placed on a microscope slide, lysis by detergents and high salt, and electrophoresis under neutral conditions (pH 7.5), for the detection of double stand breaks. In 1988, Singh *et al.* introduced the microgel technique performed under alkaline conditions (pH >13), which was capable of detecting DNA single strand breaks and double stand breaks (low levels) resulting from direct interaction with the DNA or formed as a consequence of DNA repair, and alkali-labile sites created by DNA

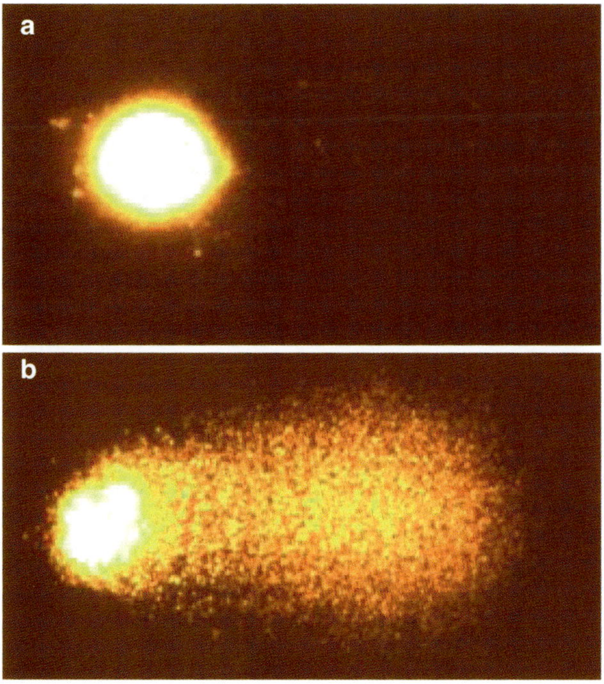

Fig. 5.1 Images of comets formed by TK6 cells with no DNA damage (**a**) and with DNA damage after methyl methanesulfonate (MMS) exposure (**b**), manifested by an increase in migration of DNA fragments outside the nucleus (*comet shape*)

damage [14–17]. Since the majority of the genotoxic agents induce more single strand breaks and alkali labile lesions compared to double strand breaks, the alkaline assay offers increased sensitivity for detecting induced damage than the neutral version of the assay [17]. Whereas the alkaline elution assay has been historically used for detecting DNA damage [18], the Comet assay provides important advantages that include: the requirement for only a small number of cells, the ability to measure damage at the single cell level, and compatibility with acute and subchronic toxicology study designs. In addition, the *in vivo* alkaline Comet assay can provide valuable information on different tissues which makes it a good complement to assays that detect genotoxicity in hematopoietic cells [19,20].

Technical variables can affect the sensitivity and resolving power of the Comet assay. It is important to establish and rigorously maintain consistent experimental conditions for the set of assays and samples (*i.e.,* tissues) to be analyzed [17,20]. The major technical variables in the Comet assay are the unwinding time, composition and pH of the electrophoresis buffer, and the electrophoresis conditions, such as voltage (V/cm) and duration [19,21]. Cells with DNA damage, *i.e.*, having increased strand breaks, display increased migration of the fragmented DNA toward the anode following electrophoresis, forming a 'tail' from the nucleus which results in a structure with the appearance of a comet (Fig. 5.1). The magnitude of DNA damage can be quantitated based on the amount of fragmented DNA found in the comet tail (referred to as tail intensity). Tail intensity is usually measured using an automatic (or semi-automatic) image analysis system that can detect fluorescently stained DNA (*i.e.,* stained with SYBR gold, Syber Green, propidium iodide, ethidium bromide, etc.) [22].

The Comet assay is widely used for environmental monitoring, human biomonitoring, molecular epidemiology, and fundamental research on DNA damage and repair [23–25]. In the pharmaceutical industry, both *in vitro* and *in vivo* versions of the Comet assay are used for genetic toxicology testing. *In vitro* testing has been used mainly for screening purposes or for mechanistic assessments, whereas, the *in vivo* Comet assay has become one of the recommended *in vivo* assay options for the genotoxicity testing required by regulatory agencies for the registration of pharmaceuticals [11]. A major advantage of the *in vivo* alkaline Comet assay is that it can be performed with almost any animal tissue, as long as a single cell suspension can be obtained with minimal cell damage, thereby providing mechanistic and/or target organ specific toxicity information [20,26].

A formal validation trial for the *in vivo* alkaline Comet assay was led by the Japanese Center for the Validation of Alternative Methods (JaCVAM) from 2006 to 2012, which paved the way for the adoption of OECD Test Guideline (TG) 489 (*In vivo* mammalian alkaline Comet assay) [27,28]. In brief, the study design outlined in this TG is divided into two phases, a dose range-finding study to determine the maximum tolerated dose (MTD), and a definitive study conducted in rodents. The design of the definitive assay includes: (1) five animals per group/sex, except for the positive control group where three animals/sex are acceptable; (2) five dose groups consisting of a concurrent vehicle control, a positive control, and three test article dose levels, where the high dose is the MTD or 2000 mg/kg/day; (3) a route of administration that mimics the anticipated route of human exposure; and (4) an exposure schedule that can vary from one, two or multiple treatments, depending on the properties of the chemical to be tested. Whatever treatment schedule is used, the tissue harvest time is critical, and should reflect the period needed for the test compound to reach a maximum concentration in the target or surrogate tissue, and for DNA strand breaks to be induced. Ideally, the sampling time(s) should be determined from kinetic data (e.g., the time at which the peak plasma concentration is achieved (Tmax)). In the absence of kinetic data, the tissues should be sampled 2–6 h after the last treatment for two or more daily treatments, or at both 2–6 and 16–26 h after a single administration.

One important feature of OECD TG 489 is the requirement that the testing laboratory show proof of proficiency in running the assay, for each tissue that is analyzed. In addition, the Comet assay can be integrated with other toxicological studies, *i.e.* integrating the *in vivo* mammalian erythrocyte MN assay and the Comet assay into a repeat dose toxicity study) to make maximum use of animal resources [28–30].

OECD TG 489, adopted in September 2014, is anticipated to result in a broader use of the Comet assay for regulatory purposes, because the ability to test multiple and different organs will complement current genotoxicity assessments. In recent years, the pharmaceutical industry has used the *in vivo* Comet assay as a follow-up test to develop weight of evidence in assessing results of *in vitro* or *in vivo* assays. Recommendations incorporated in the new ICH S2R1 guidance mean that the pharmaceutical industry can now use the *in vivo* Comet assay as one of the primary tests in the standard genotoxicity testing battery (Option 2). A recent paper by Frötschl [31], illustrates the experiences in the German regulatory agency (BfArM) on the submission of *in vivo* Comet assay data. The majority of the *in vivo* Comet assay submissions were to follow up *in vivo* relevance of *in vitro* positive cytogenetic assays. However, this assessment came before the approval of the OECD guidelines, so the use of the *in vivo* Comet assay as the second *in vivo* assay in ICH S2R1 Option 2 may increase.

5.3 *Pig-a* Gene Mutation Assay

Methods have been developed for quantifying the frequency of hematopoietic cells that exhibit inactivating mutations in the phosphatidylinositol glycan-class A gene (*Pig-a* in rodents, *PIG-A* in humans) [32–35]. These methods are based on flow cytometric assessment of glycosylphosphatidylinositol (GPI) anchored protein expression on the cell surface. Whereas human *PIG-A* assays have focused on transformed lymphoblastoid cell lines and peripheral blood leukocytes [32,33,36], studies involving rodent models have generally used circulating erythrocytes which can be obtained in abundance *via* small volume blood draws [37,38]. Whatever the target cell population, the principle is the same—GPI anchor deficiency is a characteristic of *Pig-a* mutation, and this can be readily detected with fluorescent antibodies against GPI-anchored cell surface markers such as CD59, CD55, and/or CD24 (Fig. 5.2).

Interest in the rodent *Pig-a* erythrocyte mutation assay has been high, in part due to the ease with which it can be integrated into other studies and its complementarity to the MN endpoint. Integrating these two endpoints into routine toxicology studies and short-term multi-endpoint genetic toxicology studies provides measures of gene mutation as well as clastogenic and aneugenic activity [39–45].

International trials have been organized to investigate the merits and limitations of rodent erythrocyte *Pig-a* assays [46,47]. The promising data resulting from these efforts, coupled with the compatibility of the endpoint with integrated study designs, encouraged the Steering Committee of the International Workshop on Genotoxicity Testing (IWGT) to add the *in vivo Pig-a* assay to their list of six topics covered by their 2013 program in Brazil, and the resulting report has been published in a Special Issue of Mutation Research [48].

As noted by the IWGT workgroup, an important experimental design consideration is one shared by the *in vivo* erythrocyte-based MN assay. That is, for confidence in a negative *Pig-a* result (*i.e.,* no apparent effect), it is important to demonstrate systemic exposure to the test article. Determination of systemic exposure is required since the mutations that give rise to circulating *Pig-a* mutant phenotype erythrocytes occur in erythroid precursors, cells that are usually present only in the well perfused bone marrow compartment. While demonstration of systemic exposure is important, it should be noted that, as with the MN endpoint, this

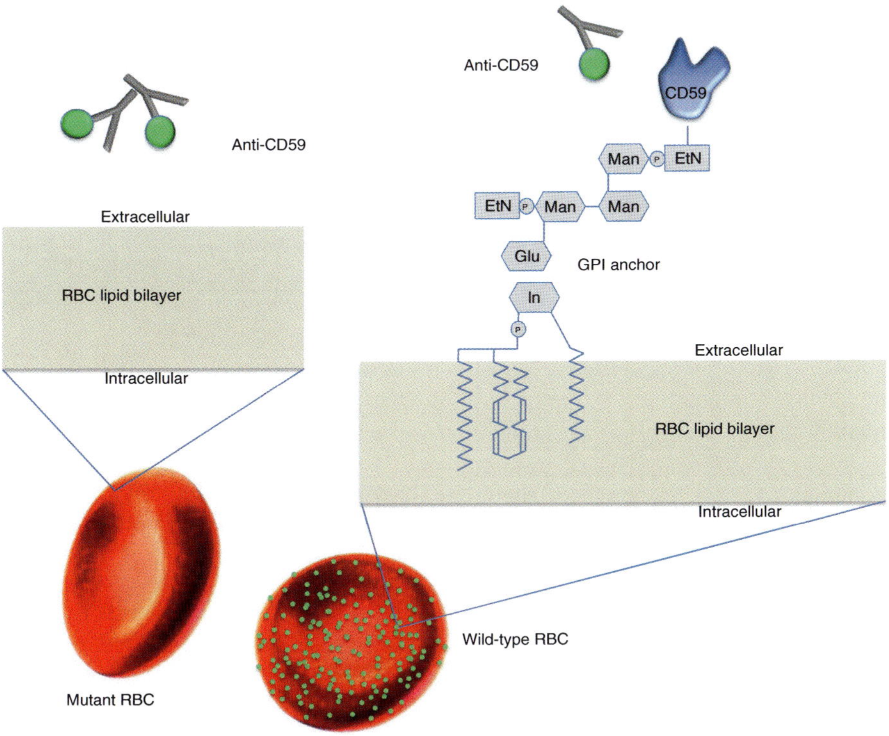

Fig. 5.2 Hematopoietic cells, including red blood cells as shown here, require GPI anchors to attach certain proteins to their cell surface (*e.g.*, CD59). Mutation of the *Pig-a* gene can prevent functional anchors from being produced, resulting in cells lacking these proteins on their surface. Fluorescent antibodies directed against GPI-anchored protein(s), in conjunction with flow cytometric analysis, thereby provide a means to score the frequency of *Pig-a* mutant phenotype cells (Red blood cells modified from Wiki Media, http://commons.wikimedia.org/wiki/File:Blausen_0761_RedBloodCells.png)

requirement does not preclude effective testing of most classes of chemicals. Evidence for the general effectiveness of using peripheral blood erythrocytes to evaluate the in vivo mutagenic potential of chemical agents comes from experiments with promutagens that are primarily bioactivated to reactive electrophiles in the liver. For instance, treatment of rats with each of the following promutagens has been observed to increase the frequency of *Pig-a* mutant erythrocytes: 2-acetylaminofluorene, aflatoxin B1, aristolochic acids, benzo[*a*]pyrene, cyclophosphamide, diethylnitrosamine, dibenzo[*a,l*]pyrene, 7,12-dimethyl-1,2-benz[*a*]anthracene, and urethane (reviewed by Gollapudi et al. [48]).

These data, in combination with international trials and other reports, prompted the IWGT workgroup to conclude that the erythrocyte-based *in vivo* rat assay is ready to play a useful and important role in the area of regulatory science, stating '…the Workgroup recognizes the assay's potential for integration with other *in vivo* tests, its potential for clinical translation, its remarkable sensitivity to mutagenic agents, its relatively low cost, and, above all, the nature of the endpoint that it measures (phenotypic mutation in an endogenous mammalian gene). The Workgroup concludes that these features strongly recommend adoption of the assay to supplement existing *in vivo* genetic toxicology assays' [48].

One area where rodent *Pig-a* assays may find use in regulatory safety assessment is as follow-up to Ames-positive drug impurities that cannot be readily controlled to an acceptable limit. For example, ICH Guideline M7 lists the *in vivo Pig-a* assay as a possible follow-up test for Ames-positive drug impurities [49]. Since recommendations for following up Ames-positive results emphasize the desirability of studying a mutational endpoint as opposed to cytogenetic damage, there will be times when the *Pig-a* assay will represent a suitable and cost-effective choice. As stated above, an essential requirement for using the assay for this and other applications is the demonstration of systemic exposure. Beyond testing genotoxic impurities, other regulatory uses are likely to become more common, especially once an OECD TG is developed that describes the minimum requirements for conducting a regulatory-compliant *in vivo Pig-a* assay. The cross-species potential of the assay is another attractive aspect of this mutation analysis platform. It is conceivable that in the future, and with additional validation, this translational potential may allow drugs with equivocal preclinical genotoxicity profiles to advance to clinical trials so long as the first studies in humans include this mutation endpoint [50,51].

5.4 Liver Micronucleus

Hamada and colleagues have developed a liver MN assay that involves the repeat dosing of rats (*e.g.*, for 28-days), the preparation of hepatocytes using *ex vivo* incubation with collagenase, and scoring of micronucleated hepatocytes by microscopy [52]. It uses a well-understood endpoint, MN induction, measured not in hematopoietic tissue, but in the metabolically active, but slowly dividing liver of adult rats. Previous methods for measuring micronuclei in liver employed acute doses and required using young rats (whose liver cells were more actively replicating) or stimulating liver cell replication in adult animals through partial hepatectomy or mitogen treatment. Also, previous methods typically used *in situ* perfusion of the liver to obtain an adequate population of cells for scoring. The repeat-dose treatment schedule appears to result in a sufficient accumulation of micronucleated hepatocytes in adult rats to detect the genotoxicity of liver-specific compounds like diethylnitrosamine, which are not easily detected in the hematopoietic MN assay [53]. Also, using only a small piece of liver to prepare cells for scoring, and processing the tissue *ex vivo*, without perfusion, leaves the remainder of the liver available for histopathological evaluation and other assays.

The assay addresses the need for measuring *in vivo* genotoxicity using a tissue other than hematopoietic tissue. It is similar in this respect to the *in vivo* Comet assay; however, this particular version of the MN assay is limited to the liver. The assay has generated considerable interest in Japan, and the Collaborative Study

Group for the Micronucleus Test, which is part of the Mammalian Mutagenesis Study group of the Environmental Mutagen Society of Japan, is expanding the number of agents tested in the assay [54]. A Workgroup at the 2013 IWGT meeting reviewed data from all versions of the rat liver MN assay, including the repeat dose version [55]. Although there have been relatively few agents tested to date, and the test appears to be insensitive to aneugens, the Workgroup found that the sensitivity and specificity of the assay for hepatocarcinogens are quite high. The Workgroup indicated, however, that there was uncertainty about how best to evaluate toxicity in the assay and that there are potential limitations in the number of cells that can be scored. On the plus side, the assay has excellent potential for integration into repeat-dose general toxicity studies.

5.5 Assays to Follow Up or Support Positive Results

Regulatory agencies and pharmaceutical companies require an understanding of the genotoxic and, by extension, the carcinogenic potential of compounds prior to their first use in clinical trials. Part of this genotoxicity assessment is done *in vitro*, using bacteria and mammalian cells. The majority of these assays have been used for more than 30 years (*i.e.,* Ames assay, chromosomal aberrations, mouse lymphoma and *in vitro* MN) and they are considered to be good hazard identification tools. While these *in vitro* genotoxicity assays have a high sensitivity for detecting DNA damage, they exhibit relatively low specificity. That is, it has been argued that they produce too many "false" or "irrelevant" positive results [9]. This is not to say the *in vitro* DNA-damaging effects did not occur in the *in vitro* assay; rather that the positive result cannot be recapitulated *in vivo*. In addition, there is a widely held view that a major culprit in producing irrelevant positive *in vitro* responses is test article-induced cytotoxicity that secondarily affects DNA [56,57].

The incidence of irrelevant positive *in vitro* assay results has led to various strategies for evaluating the significance of suspect findings. While varied, these approaches have a common goal—to deduce whether the effect(s) observed *in vitro* have *in vivo* relevance, and/or whether the mode of genotoxic action occurs through a DNA-reactive mechanism, or whether damage is secondary and related to off-target effect(s). Understanding direct versus indirect effects can be important, because it may provide a more in-depth understanding on the mechanism of action, possible effects in humans, and for certain assays, there is a possibility to develop a margin of exposure argument if sufficient differences exist between efficacious drug plasma levels and those that induce genotoxicity. Several illustrative follow-up strategies are briefly described in the following paragraphs.

As noted above, *in vitro* cytotoxicity may result in irrelevant positive findings. One approach has been to add more and novel cytotoxicity endpoints to the relatively simple characterizations of cytotoxicity that are employed in mammalian cell culture-based genotoxicity assays (*e.g.,* relative increases in cell counts, population doublings, etc.). For example, in the context of the *in vitro* Comet assay, Shi and colleagues [58] described advantages to assessing cellular adenosine triphosphate (ATP) content and activation of Caspase-3/7 instead of simple Trypan blue dye exclusion and relative cell counts. More recently Bryce *et al.* [59] described a matrix of biomarkers that provide information on cytotoxicity, double-strand DNA breaks, and the proportion of metaphase cells, to determine if *in vitro* MN formation resulted from clastogenic, aneugenic, or cytotoxic activity.

When an *in vitro* positive finding is suspected of being due to aneugenicity as opposed to direct DNA-reactivity, there are often several possible paths forward. This is because aneugenicity demonstrates thresholds below which genotoxicity is not likely to occur [60]. Approaches for demonstrating aneugenicity include evaluating chemical-induced micronuclei for the presence of kinetochores using CREST antibodies [61], and/or establishing the presence of centromeric DNA with appropriate FISH probes [62]. In both cases, such data serve to demonstrate that the micronuclei were the result of lagging chromosomes (aneugenic activity) as opposed to double-strand breaks.

A third general strategy involves using well-considered rodent-based assay(s) to follow-up *in vitro* positive results. When the questionable *in vitro* result occurred in a mutation assay, it is most preferable to study the test article's potential to cause *in vivo* mutation. When the study in question was a cytogenetic assay, it is preferable to evaluate the test article's ability to induce DNA damage *in vivo*. It is recommended that the *in vivo* assessments occur in two different tissue compartments. The bone marrow is usually tested in the context of an *in vivo* MN assay performed with bone marrow or peripheral blood erythrocytes. Careful choice of the additional tissue(s) can be important, especially when there is reason to believe the genotoxic effect may be due to a short-lived metabolite that might not become systemically available at sufficient concentration. In such cases, assays conducted in the liver are often advisable, for example the *in vivo* Comet assay, a TGR mutation assay, or the liver MN assay.

5.6 New Developments for an Old Assay: TGR

The transgenic rodent (TGR) gene mutation assay was first proposed in the early 1980s, with the first practical models, Muta™mouse and Big Blue® mouse, appearing in the late 1980s and early 1990s [63]. The assay uses bacterial transgenes, stably integrated into the genomic DNA of rodents, as reporters of *in vivo* gene mutation. To measure mutation, DNA is extracted from the tissues of interest, and the recovered transgenes evaluated for mutation in bacterial host strains. Although the first TGR assays were quite tedious (and expensive) to perform, better models employing both mice and rats, positive selection of mutations, and targets sensitive to a wider range of mutational events, have appeared over the years. Arguably, the most important development was that of a standard protocol for conducting the assay, based on the kinetics of *in vivo* gene mutation and the properties of the transgene reporters [64]. All this activity eventually led to OECD TG 488, approved in 2011 and revised in 2013 [65], making the TGR assay more attractive for regulatory testing.

The TGR assay is recommended for use in some of the latest regulatory testing guidelines for human pharmaceuticals. For instance, the ICH M7 Guideline lists the TGR assay as a possible follow-up assay for Ames-positive drug impurities that cannot be controlled at an acceptable limit [49], and Option 2 of ICH S2(R1) indicates that it could serve as companion assay to the hematopoietic MN assay [11]. Of note, the TGR assay was used for evaluating the *in vivo* dose response mutagenicity of ethylmethanesulfonate (EMS) as part of the experiments done to evaluate the risks to patients who received EMS-contaminated Viracept as a result of an industrial accident that occurred in 2007 [66].

Another area where the TGR assay may have future impact is in evaluating germ cell mutagenicity. The TGR assay is the only gene mutation assay for germ cells with an OECD TG [65]. This could mean more use of the TGR assay in the future. For example, the European Chemicals Agency's REACH (Registration, Evaluation, Authorization, and Restriction of Chemicals) guidelines require specific consideration of germ cell mutation for registering industrial and other chemicals [67]. If there are insufficient data to categorize an agent as to germ cell mutagenicity, an 'appropriate test' for germ cell mutation may have to be performed. Whether or not an agent is a germ cell mutagen can have major economic consequences. How these guidelines will be implemented is presently unknown, and it is also unclear if such considerations will impact the safety evaluation of human pharmaceuticals. Currently, however, testing labs that perform the TGR are expanding the small database for its use in detecting germ cell mutation.

5.7 New Approaches for Conducting Genetic Toxicology Testing and Using Genetic Toxicology Data

5.7.1 Combination and Integrated Studies

A paradigm shift has occurred with respect to the way *in vivo* genotoxicity assays are conducted. Whereas *in vivo* assays were traditionally performed as dedicated, one-endpoint studies, there has been a strong move towards combining multiple genotoxicity endpoints into acute studies, and also integration of two or more genetic toxicology endpoints into repeat-dose general toxicology studies. For example recent guidance on the requirements for genotoxicity testing of pharmaceuticals (ICH S2 (R1); [11]), pesticides [68], and chemicals [67], as well as OECD *in vivo* genetoxicty testing guidelines (TG 474, TG 475 and TG 489), encourage integration of genotoxicity tests into repeat-dose toxicity studies whenever possible and scientifically justified.

There are several reasons for these changes, starting with reductions in animal use. However, it must be emphasized that reduced animal use and other resource-saving advantages do not mean the assays are compromised or somehow hampered. Rather, combination of endpoints and integration into repeat-dose studies provide richer, more informative data sets. These practices allow the DNA damage assessments to be evaluated in conjunction with toxicological information obtained in the same study, which often include exposure data, hematology, clinical chemistry, and histopathology. These types of information are invaluable for interpreting in vivo genotoxicity results, whether the findings are positive or negative.

As noted in the Introduction, no single genotoxicity endpoint covers every type of DNA damage that is important for hazard identification and risk assessment. This is why combination and integrated studies strive to include endpoints that are responsive to a range of genotoxic activities. An increasingly common experimental design involves several consecutive days of treatment with the test article, followed by blood and tissue harvest several hours following the last exposure. In this scenario, micronucleated immature erythrocytes approach a steady-state level, so this one sampling time is adequate whether the cells are collected from bone marrow or blood. By carefully planning the harvest time several hours after last treatment, it is possible to collect Comet assay data that meet current recommendations. With this combination, multiple tissues are considered, and a broader range of genotoxic activities are investigated compared to traditional designs.

In terms of integrating genotoxicity endpoints into ongoing repeat-dose toxicology studies, certain logistical obstacles are often encountered. For instance, given the advice to collect tissues several hours after last administration, integration of the Comet assay requires an additional exposure than is ordinarily used for a 28-day study. Thus, the requirement for an extra ("Day 29") exposure requires a modification to the standard test. Furthermore, given the Comet assay's requirement for tissues to be collected and processed in a very timely manner, careful coordination and staggering of harvest times may be necessary. These are not insurmountable problems, but rather logistical considerations that must be anticipated and addressed. In some cases it may be possible to use animals used for toxicokinetic measurements, as opposed to main study animals, a strategy that simplifies many of these logistical concerns. Furthermore, it is generally recognized that certain genotoxicity endpoints are easily incorporated into on-going toxicology studies, even main study animals. Endpoints that are readily integrated include measurement of micronucleated erythrocytes and erythrocyte *Pig-a* mutant frequencies, DNA damage using blood cell-based Comet assays, and liver hepatocyte micronuclei.

It is likely that 3Rs principles, in combination with other incentives for acquiring concurrent general toxicology data along with several genotoxicity endpoints, will continue to drive improvements to the design of regulatory safety assessment studies.

5.7.2 Refining Assays for Greater Statistical Power

Many assays and approaches for quantitatively evaluating DNA damage have been developed over the past several decades. Given the fidelity of repair processes, one common theme is that unrepaired DNA damage is a relatively rare event. This makes precise enumeration of certain forms of DNA damage challenging. For

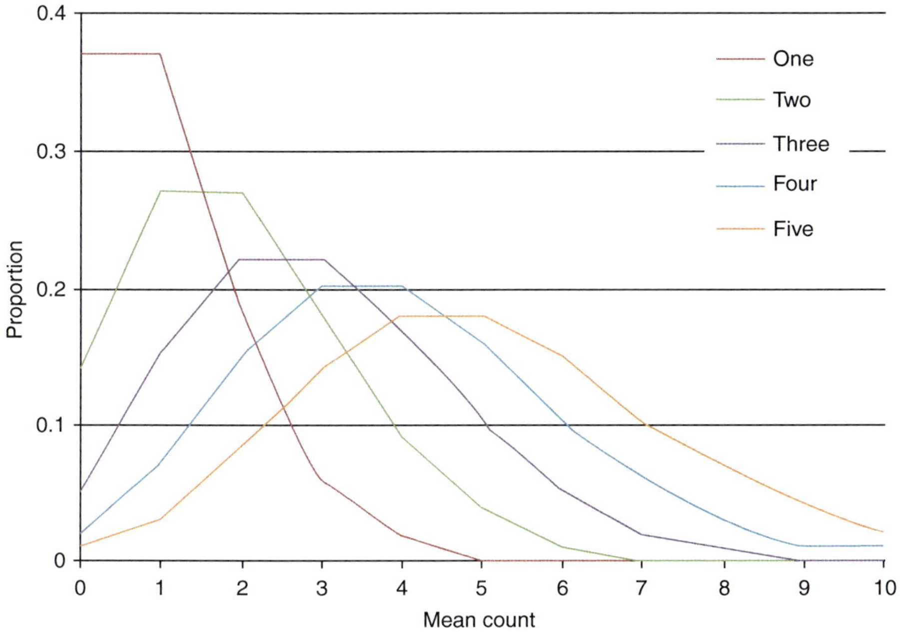

Fig. 5.3 *Y*-axis shows proportion of counts when the underlying mean of the population is one, two, three, four or five micronuclei (*X*-axis). For example, 37 % of samples will have a zero micronucleus count when the mean value is one, whereas less than 10 % of samples will have a zero count when the mean value is equal to or greater than 3 (These data assume Poisson distribution; the figure is adapted from the OECD [70], Table 10, pg 76)

example, microscopy has traditionally been employed to score the incidence of micronucleated immature erythrocytes. Given the rarity of micronucleated immature erythrocytes, typically on the order of 1–3 per 1000 cells, guidance documents had until recently advised scoring a minimum of 2000 immature erythrocytes per animal. In part, this reflected practical considerations associated with the tedious and time-consuming nature of rare event scoring *via* microscopy. However, an updated OECD TG 474 was developed with a new emphasis on increasing the statistical power of the assay [69]. Thus, in the revised guideline, the new recommendation is to score at least 4000 immature erythrocytes per animal for the presence of MN. As described in OECD report No. 198 [70], this has the effect of reducing the number of samples with MN counts of zero, a situation that contributes to assay power (Fig. 5.3). Whereas improvements to statistical power also could be achieved by increasing the numbers of animals per treatment group, this was viewed as contrary to 3Rs principles. Thus, the requirement for greater numbers of cells per animal represented a means for ensuring greater confidence in a negative result, without increasing animal numbers, but rather at the expense of more time evaluating cells for the presence of micronuclei. It should be noted that solutions exist for reducing data acquisition times. For example, regulatory-accepted flow cytometric scoring methods are now available, and these tools facilitate evaluation of 4000 or more immature erythrocytes for the presence of micronuclei in shorter periods of time and with greater objectivity than is possible with microscopy [71].

The *Pig-a* assay represents an even more extreme example of rare event scoring that requires careful consideration of statistical power. For this endpoint, typical spontaneous mutant erythrocytes frequencies are on the order of several per million total erythrocytes [37]. While early experiments that evaluated less than or equal to one million cells per rodent were sufficiently sensitive to detect marked induction following exposure to potent mutagens, power analyses indicated that more typical mutagens would often be missed unless experiments utilized unacceptably larger numbers of animals per treatment group. The solution was to develop more advanced scoring methodology that made it feasible to score many millions of cells per rodent in a short amount of time (minutes). This was accomplished by adding immunomagnetic separation technology to the flow cytometry-based evaluations [72,73]. Micronuclei and *Pig-a* scoring therefore serve as examples of regulatory genetic toxicology's greater emphasis on statistical power as opposed to ease and convenience, a situation that ultimately provides in greater confidence in negative results.

5.7.3 Mutation as an Apical Endpoint and Quantitative Approaches

Perhaps the most radical recent development in genetic toxicology involves rethinking the paradigm by which genetic toxicology data are used for regulatory safety assessments. First, since mutation is involved in many human diseases, not only cancer (e.g., [74]), mutation should be the 'apical' endpoint for genotoxicity testing [75]. This harkens back to the time when germ cell mutation *per se* was considered an apical endpoint and not its relationship to a particular genetic disease [76]. Mutation is involved in cancer but mutation ≠ cancer; mutation however is a risk factor for disease and should be regulated in its own right. One idea is to consider mutation, along with other toxicology data, to establish putative Adverse Outcome Pathways (AOPs). The AOP(s) for a test agent may or may not result in cancer and may or may not require further support by additional genetic toxicology testing.

Second, it should be acknowledged that many substances can induce mutation at a high enough dose, again proving the wisdom of the adage ascribed to Paracelsus (1493–1541): 'The dose makes the poison'. It also is the case that the dose responses for many genotoxic substances are not linear, and that at low enough doses it is difficult to detect a genotoxic response. There are also mechanisms, like DNA excision repair and Phase II conjugation reactions, which imply that no-effect levels of genotoxins are not only plausible but an expected consequence of biology, and not necessarily a lack of sensitivity of assay systems. Whereas the interpretation of dose repose data is central to most toxicological evaluations, genetic toxicology data historically have been used to divide substances into those that are positive and those that are negative, and those that are positive are often positive only at relatively high doses. There is some consideration of dose in the M7 guidance, where low enough levels of some drug impurities, even highly mutagenic and carcinogenic substances, are considered virtually safe and acceptable as impurities in drug preparations. Also the European Medicines Agency concluded that the relatively high level of EMS contamination in Viracept (see above) posed no risk to humans because experimental evidence (including TGR data) indicated that the NOGEL for EMS genotoxicity sufficiently exceeded the concentrations to which humans were exposed.

These are emerging ideas and are not sufficiently developed to cause a wholesale rethinking of present regulatory approaches. However, the Health and Environmental Sciences Institute Genetic Toxicology Technical Committee is considering several questions regarding the application of quantitative metrics to genetic toxicology data. Unknowns include which tests are best suited for dose response modeling, what is the best way to conduct the modeling and how dose response data should be interpreted in terms of human risk [75,77]. One idea is to use the dose-response data to estimate Points of Departure (PoDs), like Bench Mark Doses or Slope Transition Doses, from different types of dose response data. The PoDs can then be used to establish a virtually safe dose, e.g., a Permitted Daily Exposure or PDE, as is done with other types of toxicology data. Preliminary analyses conducted using these approaches are encouraging [75].

5.8 Conclusions

The discipline of genetic toxicology continues to advance with the adoption of new assays, scoring methods, and paradigms for interpreting results. These new tools and approaches are expected to contribute to better risk assessments and regulatory safety decisions.

References

1. Zeiger E (2004) History and rationale of genetic toxicity testing: an impersonal, and sometimes personal, view. Environ Mol Mutagenesis 44:363–371
2. Zeiger E (2010) Historical perspective on the development of the genetic toxicity test battery in the United States. Environ Mol Mutagenesis 51:781–791
3. Ames BN, Lee FD, Durston WE (1973) An improved bacterial test system for the detection and classification of mutagens and carcinogens. Proc Natl Acad Sci U S A 70:782–786
4. Ames BN, Durston WE, Yamasaki E, Lee FD (1973) Carcinoges are mutagens: a simple test system combining liver homogenates for activation and bacteria for detection. Proc Natl Acad Sci U S A 70:2281–2285
5. Cimino MC (2006) Comparative overview of current international strategies and guidelines for genetic toxicology testing for regulatory purposes. Environ Mol Mutagenesis 47:362–390
6. Casciano DA (1991) Introduction: historical perspectives of genetic toxicology. In: Li AP, Heflich RH (eds) Genetic toxicology. CRC Press, Boca Raton, pp 1–12

7. Müller L, Kikuchi Y, Probst G, Schechtman L, Shimada H, Sofuni T, Tweats D (1999) ICH-harmonised guidances on genotoxicity testing of pharmaceuticals: evolution, reasoning and impact. Mutat Res 436:195–225
8. Tennant RW, Zeiger E (1993) Genetic toxicology: current status of methods of carcinogen identification. Environ Health Perspect 100:307–315
9. Kirkland D, Aardema M, Henderson L, Müller L (2005) Evaluation of the ability of a battery of three in vitro genotoxicity tests to discriminate rodent carcinogens and non-carcinogens I. Sensitivity, specificity and relative predictivity. Mutat Res 584:1–256
10. Parsons BN (2008) Many different tumor types have polyclonal tumor origin: evidence and implications. Mutat Res 659:232–247
11. ICH (International Conference on Harmonisation of Technical Requirements for Registration of Pharmaceuticals for Human Use). (2011) Guidance on genotoxicity testing and data interpretation for pharmaceuticals intended for human use, S2(R1). Available at: http://www.ich.org/fileadmin/Public_Web_Site/ICH_Products/Guidelines/Safety/S2_R1/Step4/S2R1_Step4.pdf
12. Rothfuss A, Honma M, Czich A, Aardema MJ, Burlinson B, Galloway S, Hamada S, Kirkland D, Heflich RH, Howe J, Nakajima M, O'Donovan M, Plappert-Helbig L, Priestley C, Recio L, Schuler M, Uno Y, Martus H-J (2011) Improvement of in vivo genotoxicity assessment: combination of acute tests and integration into standard toxicity testing. Mutat Res 723:108–120
13. Ostling O, Johanson KJ (1984) Microelectrophoretic study of radiation-induced DNA damages in individual mammalian cells. Biochem Biophys Res Commun 123:291–298
14. Singh NP, McCoy MT, Tice RR, Schnider EL (1988) A simple technique for quantitation of low levels of DNA damage in individual cells. Exp Cell Res 175:184–191
15. McKelvey-Martin VJ et al (1993) The single cell gel electrophoresis assay (comet assay): a European review. Mut Res 288(1):47–63
16. Fairbairn DW, Olive PL, O'Neill K (1995) The comet assay: a comprehensive review. Mutat Res 339:37–59
17. Tice RR (1995) The Single Cell Gel/Comet Assay. Environmental Mutagenesis. In: Phillips DH, Venitt S (eds) Environmental mutagenesis. BIOS Scientific Publishers, Oxford, UK, pp 315–339
18. Parodi S, Taningher M, Santi L, Cavanna M, Sciaba L, Maura A, Brambilla G (1978) A practical procedure for testing DNA damage in vivo, proposed for a pre-screening of chemical carcinogens. Mut Res 54:39–46
19. Tice RR, Arguello E, Anderson D, Burlison B, Hartman A, Kobayashi H, Miyamae Y, Rojas E, Ryu JC, Sasaki F (2000) Single cell gel/comet assay:guidelines for invitro and in vivo genetic toxicology testing. Environ Mol Mutagen 35:206–221
20. Hartmann A et al (2003) Recommendations for conducting the in vivo alkaline Comet assay. Mutagenesis 18(1):45–51
21. Guérard M, Marchand C, Plappert-Helbig U (2014) Influence of experimental conditions on data variability in the liver comet assay. Environ Mol Mutagen 55(2):114–121
22. Kumaravel TS, Jha AN (2006) Reliable Comet assay measurements for detecting DNA damage induced by ionizing radiation and chemicals. Mut Res Genet Toxicol Environ Mutagen 605(1–2):7–16
23. Escobar PA, Olivero OA, Wade NA, Abrams EJ, Nesel CJ, Ness RB, Day RD, Day BW, Meng Q, O'Neill JP, Walker DM, Poirier MC, Walker VE, Bigbee WL (2007) Genotoxicity assessed by the comet and GPA assays following in vitro exposure of human lymphoblastoid cells (H9) or perinatal exposure of mother-child pairs to AZT or AZT-3TC. Environ Mol Mutagen 48(3–4):330–343
24. Escobar PA, Smith MT, Vasishta A, Hubbard AE, Zhang L (2007) Leukaemia-specific chromosome damage detected by comet with fluorescence in situ hybridization (comet-FISH). Mutagenesis 22(5):321–327
25. Collins AR (2004) The Comet assay for DNA damage and repair: principles, applications, and limitations. Mol Biotechnol 26(3):249–261
26. Burlinson B et al (2007) In Vivo Comet Assay Workgroup, part of the Fourth International Workgroup on Genotoxicity Testing. Fourth International Workgroup on Genotoxicity testing:

results of the in vivo Comet assay workgroup. Mut Res/Gen Toxicol Environ Mutagen 627(1):31–35

27. OECD (Organization for Economic Cooperation and Development) (2014) Reports of the JaCVAM initiative international pre-validation and validation studies of the in vivo rodent alkaline comet assay for the detection of genotoxic carcinogens, Series on testing and assessment, nos. 195 and 196. OECD Publishing, Paris
28. OECD (Organization for Economic Cooperation and Development) (2014) Guideline for the testing of chemicals; in vivo mammalian alkaline Comet Assay, TG 489. OECD Publishing, Paris
29. Rothfuss A et al (2010) Collaborative study on fifteen compounds in the rat-liver Comet assay integrated into 2-and 4-week repeat-dose studies. Mutation Res 702(1):40–69
30. Speit G, Kojima H, Burlinson B, Collins AR, Kasper P, Plappert-Helbig U, Uno Y, Vasquez M, Beevers C, De Boeck M, Escobar PA, Kitamoto S, Pant K, Pfuhler S, Tanaka J, Levy DD (2015) Critical issues with the in vivo comet assay: a report of the comet assay working group in the 6th International Workshop on Genotoxicity Testing (IWGT). Mut Res 783: 6–12
31. Frötschl R (2015) Experiences with the in vivo and in vitro comet assay in regulatory testing. Mutagenesis 30(1):51–57
32. Araten DJ, Nafa K, Pakdeesuwan K, Luzzatto L (1999) Clonal populations of hematopoietic cells with paroxysmal nocturnal hemoglobinuria genotype and phenotype are present in normal individuals. Proc Natl Acad Sci U S A 96:5209–5214
33. Chen R, Eshleman JR, Brodsky RA, Medof ME (2001) Glycophosphatidylinositol-anchored protein deficiency as a marker of mutator phenotypes in cancer. Cancer Res 61:654–658
34. Bryce SM, Bemis JC, Dertinger SD (2008) In vivo mutation assay based on the endogenous *Pig-a* locus. Environ Mol Mutagen 49:256–264
35. Miura D, Dobrovolsky VN, Kasahara Y, Katsuura Y, Heflich RH (2008) Development of an in vivo gene mutation assay using the endogenous *Pig-A* gene: I. Flow cytometric detection of CD59-negative peripheral red blood cells and CD48-negative spleen T-cells from the rat. Environ Mol Mutagen 49:614–621
36. McDiarmid MA, Albertini RJ, Tucker JD, Vacek PM, Carter EW, Bakhmutsky MV, Oliver MS, Engelhardt SM, Squibb KS (2011) Measures of genotoxicity in Gulf War I veterans exposed to depleted uranium. Environ Mol Mutagen 52:569–581
37. Dobrovolsky VN, Miura D, Heflich RH, Dertinger SD (2010) The in vivo *Pig-a* gene mutation assay, a potential tool for regulatory safety assessment. Environ Mol Mutagen 51:825–835
38. Peruzzi B, Araten DJ, Notaro R, Luzzatto L (2010) The use of PIG-A as a sentinel gene for the study of the somatic mutation rate and mutagenic agents in vivo. Mutat Res 705:3–10
39. Dertinger SD, Phonethepswath S, Franklin D, Weller P, Torous DK, Bryce SM, Avlasevich S, Bemis JC, Hyrien O, Palis J, MacGregor JT (2010) Integration of mutation and chromosomal damage endpoints into 28-day repeat dose toxicology studies. Toxicol Sci 115:401–411
40. Bhalli JA, Shaddock JG, Pearce MG, Dobrovolsky VN, Cao X, Heflich RH, Vohr H-W (2011) Report on Stage III *Pig-a* mutation assays using benzo[a]pyrene. Environ Mol Mutagen 52:731–737
41. Cammerer Z, Bhalli JA, Cao X, Coffing SL, Dickinson D, Dobo KL, Dobrovolsky VN, Engel M, Fiedler RD, Gunther WC, Heflich RH, Pearce MG, Shaddock JG, Shutsky T, Thiffeault CJ, Schuler M (2011) Report on Stage III *Pig-a* mutation assays using *N*-ethyl-*N*-nitrosourea—Comparison with other in vivo genotoxicity endpoints. Environ Mol Mutagen 52:721–730
42. Shi J, Krsmanovic L, Bruce S, Kelly T, Paranjpe M, Szabo K, Arevalo M, Atta-Safoh S, Debelie F, Klug LaForce M, Sly J, Springer S (2011) Assessment of genotoxicity induced by 7,12-dimethylbenz(a)anthracene or diethylnitrosamine in the *Pig-a*, micronucleus and comet assays integrated into 28-day repeat dose studies. Environ Mol Mutagen 52:711–720
43. Stankowski LF Jr, Roberts DJ, Chen H, Lawlor T, McKeon M, Murli H, Thakur A, Xu Y (2011) Integration of *Pig-a*, micronucleus, chromosome aberration, and comet assay endpoints in a 28-day rodent toxicity study with 4-nitroquinoline-1-oxide (4NQO). Environ Mol Mutagen 52:738–747

44. Lynch AM, Giddings A, Custer L, Gleason C, Henwood A, Aylott M, Kenny J (2011) International *Pig-a* gene mutation assay trial (stage 3): results with *N*-methyl-*N*-nitrosourea. Environ Mol Mutagen 52:699–710
45. Lemieux CL, Douglas GR, Gingerich J, Phonethepswath S, Dertinger SD, Phillips DH, Arlt VM, White PA (2011) Simultaneous measurement of benzo[a]pyrene-induced *Pig-a* and *lacZ* mutations, micronuclei and DNA adducts in Muta™Mouse. Environ Mol Mutagen 52:756–765
46. Dertinger SD, Phonethepswath S, Weller P, Nicolette J, Murray J, Sonders P, Vohr H-W, Shi J, Krsmanovic L, Gleason C, Custer L, Henwood A, Sweder K, Stankowski LF Jr, Roberts DJ, Giddings A, Kenny J, Lynch AM, Defrain C, Nesslany F, van der Leede B-JM, Van Doninck T, Schuermans A, Tanaka K, Hiwata Y, Tajima O, Wilde E, Elajouji A, Gunther WC, Thiffeault CJ, Shutsky TJ, Fiedler RD, Kimoto T, Bhalli JA, Heflich RH, MacGregor JT (2011) International *Pig-a* gene mutation assay trial: evaluation of transferability across fourteen laboratories. Environ Mol Mutagen 52:690–698
47. Kimoto T, Horibata K, Chikura S, Hashimoto K, Itoh S, Sanada H, Muto S, Uno Y, Yamada M, Honma M (2013) Interlaboratory trial of the rat *Pig-a* mutation assay using an erythroid marker HIS49 antibody. Mutat Res 755:126–134
48. Gollapudi BB, Lynch AM, Heflich RH, Dertinger SD, Dobrovolsky VN, Froetschl R, Horibata K, Kenyon MO, Kimoto T, Lovell DP, Stankowski LF Jr, White PA, Witt KL, Tanir JY (2015) The *in vivo Pig-a* assay: a report of the International Workshop on Genotoxicity Testing (IWGT) Workgroup. Mutat Res 783:23–35
49. ICH (International Conference on Harmonisation of Technical Requirements for Registration of Pharmaceuticals for Human Use) (2014) Assessment and control of DNA reactive (mutagenic) impurities in pharmaceuticals to limit potential carcinogenic risk, M7. Available at: http://www.ich.org/fileadmin/Public_Web_Site/ICH_Products/Guidelines/Multidisciplinary/M7/M7_Step_4.pdf
50. Dobrovolsky VN, Elespuru RK, Bigger CA, Robison TW, Heflich RH (2011) Monitoring humans for somatic mutation in the endogenous *PIG-A* gene using red blood cells. Environ Mol Mutagen 52:784–794
51. Dertinger SD, Avlasevich SL, Bemis JC, Chen Y, MacGregor JT (2015) Human Erythrocyte *PIG-A* Assay: an easily monitored index of gene mutation requiring low volume blood samples. Environ Mol Mutagen 56:366–377
52. Narumi K, Ashizawa K, Takashima R, Takaswa H, Katayama S, Tsuzuki Y, Tatemoto H, Morita T, Hayashi M, Hamada S (2012) lDevelopment of a repeated-dose liver micronucleus assay using adult rats: an investigation of diethylnitrosamine and 2,4-diaminotoluene. Mutat Res 747:234–239
53. Narumi K, Ashizawa K, Fijiishi Y, Tochinai R, Okada E, Tsuzuki Y, Tatemoto H, Hamada S, Kaneko K, Ohyama W (2013) Persistence and accumulation of micronucleated hepatocytes in liver of rats after repeated administration of diethylnitrosamine. Mutat Res 755:100–107
54. Hayashi A, Kosaka M, Kimura A, Wako Y, Kawasako K, Hamada S (2014) Evaluation of the repeated-dose liver micronucleus assay using N-nitrosomorpholine in young adult rats: report on collaborative study by the Collaborative Study Group for the Micronucleus Test (CSGMT)/Japanese Environmental Mutagen Society (JEMS)-Mammalian Mutagenicity Study (MMS) Group. Mutat Res. doi:10.1016/j.mrgentox.2014.04.002
55. Uno Y, Morita T, Luijten M, Beevers C, Hamada S, Itoh S, Ohyama W, Takasawa H (2015) Recommended protocols for the liver micronucleus test: report of the IWGT working group. Mutat Res 783:13–18. Available at: http://dx.doi.org/10.1016/j.mrgentox.2014.10.010
56. Bradley MO, Taylor VI, Armstrong MJ, Galloway SM (1987) Relationships among cytotoxicity, lysosomal breakdown, chromosome aberrations, and DNA double strand breaks. Mutat Res 189:69–79
57. Hilliard CA, Armstrong MJ, Bradt CI, Hill RB, Greenwood SK, Galloway SM (1998) Chromosome aberrations in vitro related to cytotoxicity of nonmutagenic chemicals and metabolic poisons. Environ Mol Mutagen 31:316–326
58. Shi J, Springer S, Escobar P (2010) Coupling cytotoxicity biomarkers with DNA damage assessment in TK6 human lymphoblastoid cells. Mutat Res 696:167–178

59. Bryce SM, Bemis JC, Mereness JA, Spellman RA, Moss J, Dickinson D, Schuler MJ, Dertinger SD (2014) Interpreting in vitro micronucleus positive results: simple biomarker matrix discriminates clastogens, aneugens, and misleading positive agents. Environ Mol Mutagen 55:542–555
60. Elhajouji A, Tibaldi F, Kirsch-Volders M (1997) Indications for thesholds of chromosome non-disjunction versus chromosome lagging induced by spindle inhibitors *in vitro* in human lymphocytes. Mutagenesis 12:133–140
61. Lynch AM, Parry JM (1993) The cytochalasin-B micronucleus/kinetochore assay in vitro: studies with 10 suspected aneugens. Mutat Res 287:71–86
62. Doherty AT, Ellard S, Parry EM, Parry JM (1996) A study of the aneugenic activity of trichlorfon detected by centromere-specific probes in human lymphoblastoid cell lines. Mutat Res 372:221–231
63. Boverhof DR, Chamberlain MP, Elcombe CR, Gonzalez FJ, Heflich RG, Gernandez LG, Jacobs AC, Jacobson-Dram D, Luijten M, Maggi A, Manjanatha MG, van Benthem J, Gollapudi BB (2011) Transgenic animal models in toxicology: historical perspectives and future outlook. Toxicol Sci 121:207–233
64. Thybaud V, Dean S, Nohmi T, de Boer J, Douglas GR, Dlickman BW, Gorelick NJ, Heddle JA, Heflich RH, Lambert I, Hartus H-J, Mirsalis JC, Suzuki T, Yajima N (2003) In vivo transgenic mutation assays. Mutat Res 540:141–151
65. OECD (Organization for Economic Cooperation and Development) (2013) Transgenic rodent smonatic and germ cell gieme mutation assay, TG 488. Available at: http://www.oecd-ilibrary.org/docserver/download/9713251e.pdf?expires=1413692829&id=id&accname=guest&checksum=2A63A91187AAFD447BE959F01F8A81D0
66. Müller L, Gocke E, Lavé T, Pfister T (2009) Ethyl methanesulfonate toxicity in Viracept-A comprehensive human risk assessment based on threshold data for genotoxicity. Toxicol Lett 190:317–329
67. ECHA European Chemicals Authority) (2014) Guidance on information requirements and chemical safety assessment, chapter R.7a: endpoint specific guidance. Available at: http://echa.europa.eu/documents/10162/13632/information_requirements_r7a_en.pdf
68. EPA (U.S. Environmental Protection Agency) (2012) Advances in genetic toxicology and integration of in vivo testing into standard repeat dose studies. Available at: http://www.epa.gov/opp00001/science/integrating-gentox-studies.html
69. OECD (Organization for Economic Cooperation and Development) (2014) OECD guideline for the testing of chemicals: mammalian erythrocyte micronucleus test, TG 474. OECD Publishing, Paris
70. OECD (Organization for Economic Cooperation and Development) (2014) Series on testing and assessment, no. 198: report on statistical issues related to OECD test guidelines (TGs) on genotoxicity. OECD Publishing, Paris
71. Dertinger SD, Camphausen K, MacGregor JT, Bishop ME, Torous DT, Avlasevich S, Cairns S, Tometsko CR, Menard C, Muanza T, Chen Y, Miller RK, Cederbrant K, Sandelin K, Pontén I, Bolcsfoldi G (2004) Three-color labeling method for flow cytometric measurement of cytogenetic damage in rodent and human blood. Environ Mol Mutagen 44:427–435
72. Dertinger SD, Bryce SM, Phonethepswath S, Avlasevich SL (2011) When pigs fly: immunomagnetic separation facilitates rapid determination of *Pig-a* mutant frequency by flow cytometric analysis. Mutat Res 721:163–170
73. Kimoto T, Chikura S, Suzuki K, Kobayashi X, Itano Y, Horibata K, Honma M, Dobrovolsky VN, Heflich RH, Miura D, Kasahara Y (2011) Further development of the rat Pig-a mutation assay: measuring rat *Pig-a* mutant bone marrow erythroids and a high throughput assay for mutant peripheral blood reticulocytes. Environ Mol Mutagen 52:774–783
74. Erickson RP (2010) Somatic gene mutation and human disease other than cancer: an update. Mutat Res 705:96–106
75. Johnson GE, Soeteman-Hernandez LG, Gollapudi BB, Bodger OG, Dearfield KL, Heflich RH, Hixon JG, Lovell DP, MacGregor JT, Pottenger LH, Thompson CM, Abraham L, Thybaud V, Tanir JY, Zeiger E, van Benthem J, White PA (2014) Derivation of Point of Departure (PoD)

estimates in genetic toxicology studies and their potential applications in risk assessment. Environ Mol Mutagenesis 55:609–623

76. DHEW (U.S. Department of Health, Education, and Welfare) (1969) Report of the Secretary's Commission on Pesticides and Their Relationshipto Environmental Health, Parts 1 and 2. U.S. GPO, Washington, DC, pp 1–677. Available at: http://eric.ed.gov/?id=ED045399
77. Gollapudi BB, Johnson GE, Hernandez LG, Pottenger LH, Dearfield KL, Jeffrey AM, Julien E, Kim JH, Lovell DP, MacGregor JT, Moore MM, van Benthem J, White PA, Zeiger E, Thybaud V (2013) Quantitative approaches for assessing dose-response relationships in genetic toxicology studies. Environ Mol Mutagen 54:8–18

Chapter 6
History and Current Regulatory Requirements

James S. MacDonald and David Jacobson-Kram

Abstract The assessment of risk to humans of cancer following exposure to chemicals has been a challenging process for decades. The early pragmatic approaches to this important challenge have evolved with growing understanding of the underlying biology of the cellular processes that lead to tumor development in animals and the relevance of these findings to human risk. The regulatory approaches to assessment of human risk of cancer in place today reflect the current state of understanding of these complex biological processes while providing a common regulatory framework for risk assessment. This chapter reviews the evolution of this process from the early days to the current state setting the framework for further evolution of how we address this critical challenge.

Keywords Bioassay history • FDA CAC • Rodent bioassay • Human risk assessment • ICH carcinogenicity guidelines • HESI Alternatives to Carcinogenicity Testing

There can be no question that exposure to some chemicals can result in the formation of tumors in humans. Since the early observations by Percival Pott of scrotal and nasal cancer in chimney sweeps in England in the 1700s [27] it has become well accepted that chemical exposure can, under certain conditions, result in human cancer. The challenge posed by this observation (and many others in subsequent years) has been and continues to be how to predict prospectively which chemicals pose carcinogenic hazards to humans under the conditions of use. This challenge takes on various dimensions as one considers occupational or environmental exposures versus those posed by deliberate application of chemicals in the context of therapeutic approach to human diseases. In the latter case, generally much more information is available on the nature of the chemical and how it interacts with biological

J.S. MacDonald (✉)
Synergy Partners R&D Solutions, LLC, 385 Route 24, Suite 1G, Chester, NJ 07930, USA
e-mail: jmacdonald@chrysalispharma.com

D. Jacobson-Kram
ToxRox Consulting, McLean, VA, USA

M.J. Graziano, D. Jacobson-Kram (eds.), *Genotoxicity and Carcinogenicity Testing of Pharmaceuticals*, DOI 10.1007/978-3-319-22084-0_6

systems including extensive safety testing in the exposed human population. In addition, the exposure has a presumed positive outcome so there is the opportunity to consider a risk-benefit analysis when evaluating the body of data. This chapter will limit the discussion of this broad topic to the subset of pharmaceutical chemicals and the particular challenges posed by these chemicals.

6.1 Early History

Research in the mid-1960s with classes of chemicals known to be carcinogenic to humans demonstrated that a similar tumorigenic response could be reproduced in animals after topical or oral administration [24, 27]. These chemicals were from classes of chemicals that are now well known to represent highly reactive and largely genotoxic chemicals such as polycyclic aromatic hydrocarbons, azo and acridine dyes, and aromatic amines and nitrosamines ([43], 40]). This observation that human tumors could be produced in animals with the same chemicals of concern led to the general concept that exposure of test animals to chemicals could predict human cancer. As greater general concern over the potential for chemicals to cause cancer grew along with the awareness of the vast number of new chemicals for which little information was available, the US National Cancer Institute proposed formal procedures for evaluating the carcinogenic potential of chemicals in the early 1960s ([3], 54]). In these proposals, it was envisioned that animal tests would identify chemicals of concern which warranted further study rather than the idea that a single study or set of studies would provide the definitive determinant of the human carcinogenic potential of a chemical [3]. Despite this reasoned approach to the challenge, the desire for a quick and simple way to identify (and remove) human cancer hazards led to the development of the recommendation in 1975 that subsequently became the basis for regulatory guidance for both environmental/industrial and pharmaceutical chemicals [48]. This guidance led to the currently accepted practice of the use of 2 year studies in two rodent species in what is generally referred to as the rodent bioassay [3]. This remains the standard today for assessment of human risk of cancer from chemical exposure.

The Sontag guidelines were published by the US National Cancer Institute in 1976 [48] and quickly became the de facto standard for assessing the carcinogenic potential of chemicals in animal studies. These guidelines gave detailed descriptions of how a rodent bioassay should be conducted and found their way into federal guidelines in the ensuing years [45]. These documents clearly described the approaches to design of the 2 year rodent studies in rats and mice including the practice of using the maximum tolerated dose (MTD) to select the top dose in these studies. The principles articulated by the NCI guided how data from these studies were used. Important among these principles was the concept that chemicals that produced tumors in rodents were assumed to present human risk and that tumors produced at a high (toxic) dose were assumed to be predictive of a tumorigenic response at lower doses [6, 45].

In the almost four decades since the guidelines for the rodent bioassay were first formally published, much has been learned about the utility of using data from these assays for human risk assessment. It has become clear that what was envisioned as a simple and efficient process for identifying potential human carcinogens is much more complex than originally thought. This complexity arose from several sources. These will be explored below in an attempt to illustrate how the continuing exploration of confounding data has led to (and is still leading to) our current understanding of how best to approach the important and difficult challenge of predicting human risk of cancer.

6.2 Growing Experience: Factors Impacting the Interpretation of Rodent Bioassay Data

When the overall body of data from rodent bioassays is considered, it can be shown that there are many more positive studies in rodents than would be expected from the relatively few known human carcinogens [41, 53]. The basis for this high rate of positive outcomes can be attributed to one or more of several factors: (1) the use of a high (toxic) dose as the top dose in the study, (2) secondary mechanisms of carcinogenicity unique to the rodent, and (3) other species-specific responses.

The available data show a high degree of concordance between the magnitude of the dose used in the bioassay and the production of tumors in rodents [20, 37]. The rationale for using the highest testable dose is that this maximizes the sensitivity of the bioassay and attempts to compensate for the limited number of animals that can be exposed in the testing environment. The consequence of this approach, however, is that mechanisms that may operate only at toxic doses may trigger processes that result in rodent tumors at the end of a 2 year study [1]. This concern over the impact of such an artificial experimental construct led to the adoption of alternative means of dose selection for these 2 year studies for pharmaceutical chemicals that will be discussed in more detail below.

As more information became available particularly with pharmaceutical compounds tested in 2 year bioassays in the 1980s and 1990s, more evidence accumulated for tumorigenic responses in rodents that were not representative of what would occur under similar exposure conditions in humans. An early observation that paved the way for a desire for more detailed mechanistic studies in evaluating rodent tumor data was the observations with soterenol. The development of this drug was stopped in the early 1970s due to the development of mesovarial leiomyomas in a 2 year study in rats. Over a period of years, it was demonstrated that this response in rats was entirely mediated by excessive β_2-adrenergic receptor stimulation and that the comparable tissue in humans does not contain this essential target [31, 32] demonstrating the species specificity of this tumorigenic response.

The so-called "secondary mechanisms" of carcinogenesis have become a more widely accepted explanation for the observation of tumors in rodents for non-genotoxic compounds. Examples of these responses include mammary tumors secondary to drug-induced alterations in prolactin levels, testicular Leydig cell tumors secondary to drug-induced increases in luteinizing hormone (LH) levels, and thyroid follicular cell tumors secondary to drug-induced increases in thyroid stimulating hormone (TSH) levels [32, 38]. Similar strong evidence was developed over many years to demonstrate why drug-induced increases in serum gastrin levels resulted in ECL cell carcinoid tumors in the stomach of rats treated with proton pump inhibitors (e.g.: omeprazole) and did not predict human tumors [4, 21, 28].

In addition to these endocrine tumors, more evidence accumulated over the years of use of the bioassay to show why some positive results in the bioassay did not predict human cancer [33]. While not specifically relating to pharmaceuticals, the demonstration that the chronic irritation and proliferative response induced in the rat bladder by sodium saccharin induced a proliferative response unique to the rodent that led to tumors in this species [5] was an important contribution to the understanding of the importance of evaluating the global weight of evidence when evaluating the significance of rodent tumors in 2 year bioassays. Similarly, the demonstration of the relationship between the appearance of so-called "hydrocarbon nephropathy" in male rats and the subsequent chronic irritation, cell proliferation and tumorigenesis with agents that bound to α2 microglobulin led to the understanding today that, for agents that can be demonstrated to act by this mechanism, rodent tumors are not considered predictive of human risk [13, 30]. Yet another example of such "secondary mechanisms" are rodent liver tumors resulting from the sustained proliferative stimulus induced by CYP enzyme inducers such as phenobarbital [19, 35, 39]. An understanding of the molecular mechanisms associated with this sustained proliferation in the rodent and the marked differences in this biology from what is observed in humans under the conditions of exposure has led to the general understanding that hepatocellular tumors that arise in rodents after prolonged exposure to such agents do not predict human risk.

6.3 A Focused Search for Better Alternatives

Over the several decades of accumulated experience with the rodent bioassay, much has been learned about how chemicals cause cancer. Our understanding of this process is very different from when the rodent bioassay was conceived and instituted. We have learned much about how pharmaceuticals (as well other chemicals) produce rodent tumors and how to interpret these findings in terms of human risk [17, 18]. The continuing problem, however, has been (and remains) the fact that the rodent bioassays take approximately 3 years to complete from initiation of dosing to final statistical evaluation of tumor data and cost from three to five million US dollars at today's prices. In addition, much time can be required to perform any mechanistic studies that may be required to put any tumor findings in perspective and understand the significance for human risk. There has been, therefore, an increasingly urgent search over the past 10–15 years for alternative approaches to the challenging task of prediction of human cancer risk particularly for non-genotoxic chemicals.

These concerns, the increasing understanding of mechanisms of non-genotoxic carcinogenicity, and the growing uncertainties around how best to interpret data derived from the 2 year bioassays led to the formation of a Expert Working Group within the ICH framework in the early 1990s and drove the discussions in this group. . The EWG initiated a retrospective examination of the pharmaceutical databases in the three regulatory regions involved in ICH (FDA – US, CPMP – Europe, MHW – Japan) and provided data that showed approximately one half of the tested compounds were positive in one or both of the two species used in the 2 year bioassays [10, 42, 53]. This exercise led one group (CHMP Safety Working Party) to conclude that little additional information of importance in human risk assessment was gained from the mouse data and put the position forward that only one species was necessary to support drug registration [53]; this opinion was not shared by the members of the other two regulatory regions.

Concurrent with the ICH discussions was the growing awareness from reports in the literature that genetically modified rodents could detect signals associated with the carcinogenic response ([50]; p53 knockout mouse: [15]; Tg.rasH2 transgenic mouse: [2]; Tg.AC mouse: [29]; XPA repair deficient mouse: DeVries [14]). Of the models that had been proposed at the time leading up to ICH III in the late 1990s, four in vivo assays were brought forward for discussion within the ICH Expert Working Group: Tg.AC knockout mouse, p53 hemizygous knockout mouse, rasH2 transgenic mouse, and the XPA repair deficient mouse. Additional models considered by the group were the newborn mouse and the in vitro SHE cell transformation assay. In a landmark publication in 1997, the US FDA stated that there was sufficient information available with several of the in vivo models to employ one of these alternative assays instead of a second species in assessing the carcinogenic potential of pharmaceutical chemicals [10]. All three regulatory authorities involved in the ICH process agreed to this in the ICH S1b guideline: "Testing for the Carcinogenic Activity of Pharmaceuticals" in that same year.

As the in vivo models had not been fully characterized with pharmaceutical chemicals at the time of the signing of this ICH guideline, a group of academic, government, and industry scientists launched a large collaborative program through the International Life Sciences Institute Health and Environmental Sciences Institute (ILSI/HESI) organization to more fully understand how best to utilize these alternative models [46]. Twenty one chemicals were evaluated including known human carcinogens, known human non-carcinogens, genotoxic and non-genotoxic chemicals in five alternative models (p53 hemizygous knockout mouse, Tg.AC transgenic mouse (using both topical and oral administration), rasH2 transgenic mouse, XPA knockout and XPA/p53 knockout mouse). In addition to these genetically modified in vivo models, the chemicals were tested in the neonatal mouse and the in vitro SHE cell transformation assay [7]. With the exception of the SHE assay and the neonatal mouse where the data were not considered sufficiently informative, the results of the studies with the other models enabled them to be used with confidence in the assessment of novel pharmaceutical compounds [16, 49, 52].

Based on the outcome of the HESI program, there was general acceptance of the use of these alternative genetically modified mouse models in both industry and regulatory agencies as screening tools. While it was initially hoped that the results of studies with these genetically modified mouse models would offer important insight into mechanism, it is clear that the genetic manipulations merely serve to enhance the sensitivity of the animal to a carcinogenic stimulus without giving information useful for mode-of-action analysis or specific tissue sensitivity [8] FDA (and agencies in other regulatory regions) generally accepts the p53 knockout mouse for evaluation of the carcinogenic potential of genotoxic agents, the rasH2 transgenic mouse for genotoxic or non-genotoxic compounds, and the Tg.AC transgenic mouse only for dermal products applied topically [25]. As of 2003, 81 protocols had been submitted for approval to the FDA using genetically modified mice [25] and as of 2005, 40 reports of completed studies with these models had been submitted to FDA [23]. Most of the protocols submitted to the FDA for approval after 2004 have employed the rasH2 mouse model [23].

6.4 Current Regulatory Approach to the Design of Studies for Carcinogenic Risk Assessment for Pharmaceuticals

6.4.1 FDA's Carcinogen Assessment Committee (CAC) and Executive CAC (eCAC)

The CAC and eCAC were established in the Center for Drug Evaluation and Research's (CDER) Office of New Drugs (OND) in the late 1980s. The need for this activity was recognized by Dr. Robert Temple (currently the Deputy Director for Clinical Science) to assure a uniform approach to study design and data interpretation across CDER and it was his leadership that led to their creation. The need for such a body becomes clear when one considers that many drugs are approved for a variety of indications. For example, duloxetine is approved for the treatment of depression, anxiety, fibromyalgia and back pain. Different indications can be regulated by different review divisions within OND. Individual divisions reviewing the same carcinogenicity studies but coming to differing conclusions could lead to regulatory chaos. As a result, OND established the eCAC and the CAC to be the final arbiter for interpreting the outcomes of carcinogenicity studies.

Other national drug regulatory bodies do not have the equivalent of CDER's eCAC. However, the Japanese and European Union regulatory bodies are signatories to ICH guidelines S1, S1A, S1B and S1C. If carcinogenicity studies are performed in conformance with these guidelines, the final results should be acceptable in all regions.

The eCAC is chaired by the pharm/tox associate director, and includes Pharmacology/Toxicology Office of Drug Evaluation (ODE) associate directors, one rotating pharm/tox supervisor and an executive secretary. The committee meets on a weekly basis to consider carcinogenicity protocols before the studies are initiated and final study reports when the studies are completed. A primary reviewer from the specific reviewing division along with his or her supervisor presents the protocol or study results to the committee. In the case of a protocol for a new study, the primary reviewer assesses adequacy of proposed study design, the adequacy of the proposed doses and the selection of the transgenic mouse model if appropriate. If the committee disagrees with sponsor's proposal, the committee suggests changes such as number of animals, doses or endpoints. If committee concurs on a protocol or if changes are proposed and accepted by sponsor, the committee cannot later reject study design. In the case of final study reports, the review group also includes a biostatistician along with his or her supervisor and presents the statistical analysis of the data.

Thirty days prior to submitting a carcinogenicity protocol, the sponsor should notify the review division of their intention to submit a Special Protocol Assessment (SPA). Submission of a SPA is not a regulatory requirement; sponsors can choose to perform carcinogenicity studies without concurrence from the FDA. However, if the design, such as dose setting turns out to be flawed, FDA is not compelled to accept the studies. Supporting dose range-finding data need to be submitted prior to or with SPA. The FDA has 45 days to respond to the SPA. Once the SPA has been evaluated by the eCAC, the executive secretary will fax or email the results of the committee's deliberations to the sponsor (Carcinogenicity Study Protocol Submissions http://www.fda.gov/downloads/Drugs/GuidanceComplianceRegulatoryInformation/Guidances/ucm078924.pdf). Concurrence must be requested prior to study initiation. If the study is performed in accordance with the eCAC recommendations, the completed study cannot be rejected based on the protocol design. If the sponsor disagrees with the eCAC their comments can be sent to the review division which will forward the inquiry to the eCAC for reconsideration. If the sponsor ignores the eCAC recommendations, it assumes the risk that a flawed design could lead to the study being rejected. There is no review clock for final study evaluations other than the PDUFA deadline for NDA review. The results of the eCAC's deliberations on final studies are not forwarded to the sponsor.

6.5 Materials to Be Included with the SPA

A 90-day toxicology study in the same rodent strain, using identical methods of administration and formulation serves to determine the maximum dose selected for the carcinogenicity bioassay as well as justifies the lower doses selected. Obviously, a 26-week study is also acceptable. For a 6-month transgenic mouse study, a 28-day dose range finding study in the wild-type strain is acceptable. Sufficient metabolism data should be submitted to demonstrate the appropriateness of the species/strain selected for covering human metabolic profile. Toxicokinetic data should enable estimation of the $AUC_{0\text{-}24}$ exposure for parent drug and any major human metabolite. Clinical exposure data at steady state for both parent and major human metabolites should be provided using the maximum recommended human dose (MRHD). Protein binding comparisons between nonclinical species and human plasma should be provided. Also a summary of the genetic toxicology data. The API used in these studies should share the same impurity profile as will be found in the marketed drug.

6.6 Dose Selection

A critical issue in designing a carcinogenicity bioassay is the selection of the high dose. A number of metrics are available for making this determination (http://www.ich.org/fileadmin/Public_Web_Site/ICH_Products/Guidelines/Safety/S1C_R2/Step4/S1C_R2__Guideline.pdf). The most commonly used criterion is the maximum tolerated dose (MTD). The MTD can be defined using a dose that causes mortality. When this criterion is used, the MTD is generally one-third of the lethal dose. Decreases in body weight from a 6-month study or decreases body weight gain (3-month study) can serve as the basis for an MTD. Decrements of approximately 10 % compared to control are generally acceptable. Toxicities in target organs that are not compatible with long term survival can be used to set an MTD; for example liver cell necrosis, erosions in the stomach or renal tubule degeneration. For some drugs clinical signs can serve as the basis for MTD selection, for example a drug that induces seizures. Altered clinical pathology parameters can define an MTD, for example if a drug interferes with blood clotting. The magnitude of these changes must be such that the doses above that which defined the MTD would not be tolerated for the duration of the carcinogenicity study.

Pharmacokinetic parameters can also be used to set the top dose. An AUC in animals that is at least 25 fold higher than the AUC at the MRHD is acceptable. A limit dose of 1500 mg/kg can be used when the drug is used clinically at less than 500 mg/day and has a 10× AUC margin. To date, the FDA has not accepted these parameters for setting the high dose in transgenic carcinogenicity studies. None of the studies used to validate the transgenic models used criteria other than the MTD.

Saturation of absorption can also be used as the basis for a top dose since increasing the dose does not result in increased exposure. Perhaps the least desirable criterion is the maximum feasible dose. This can be acceptable provided the sponsor has shown a good faith effort to test formulations and/or routes of exposure that maximize drug exposure.

6.7 Frequently Encountered Issues

Some sponsors routinely include dual vehicle controls. This practice stems from the notion that a certain amount of "noise" is expected in a carcinogenicity study. Because of the large numbers of tissues that are examined, an apparent statistical increase in a tumor could occur by chance. Having two control groups reduces the chance of such a type 1 error. In practice, CDER biostatisticians combine the information from the two controls when comparing the drug-treated groups. In years past, some sponsors routinely included calorie restriction as part of carcinogenicity. This was driven by the observation at animals fed ad libidum became obese, had shorter life-spans and increased tumor frequencies. This practice is rarely seen in contemporary studies.

6.8 Protocol Changes/Early Terminations

It is not uncommon that in the course of a carcinogenicity study sponsors wish to amend the protocol to change doses or to terminate groups prior to the scheduled necropsy. If a sponsor has received concurrence from the eCAC on the study protocol, the sponsor should contact the review division holding their IND or NDA to request such changes. The review division queries eCAC members to determine if such changes are appropriate. Response times to such requests are rapid, generally within 2–3 days.

Excessive mortality early in a study may result in recommendation to reduce the dose level. There are not hard and fast rules for early study termination. However, the following criteria are frequently followed. If animal survival drops to $\leq$20 late in the study in a manner suggesting a drug related effect, dosing may be suspended. If survival falls to 15 animals at the high dose for either males or females, after study week 100, the entire gender at all doses can be sacrificed. If the high dose in either or both sexes falls to 15 animals prior to study week 100 then just the HD group can be sacrificed and the low and mid dose can continue to the end of study.

If there are two identically treated control groups, the eCAC generally recommends combining them and if the total numbers reach 20 animals all groups of that sex can be sacrificed. If there is only one control group, all dose groups of that sex should be sacrificed once the control reaches 20 animals.

6.9 Statistical Analysis of Carcinogenicity Studies

When a carcinogenicity study is completed, the sponsor submits electronic data sets in a SAS transport file for review by statisticians at the FDA (MaPP 6610.2 Responsibilities and Procedures for Statistical Review and Evaluation of Animal Carcinogenicity Studies: http://citeseerx.ist.psu.edu/viewdoc/download;jsessionid=74700B0F1B43370E4D97B07BFD921D50?doi=10.1.1.174.1708&rep=rep1&type=pdf). Such reviews include standard assessments as well as special evaluations/combinations requested by nonclinical reviewer or eCAC. The results from this report are incorporated into the nonclinical review and presented jointly to eCAC for final adjudication.

In performing statistical analyses certain tumors should be grouped across tissues, e.g. lymphoreticular and hematopoietic neoplasms. Other tumors should be combined within organs, e.g. hepatic adenoma and carcinomas [36]. To be considered statistically significant an increase in tumors must be positive by both the trend test and by pairwise analysis. Different p-values are employed for common versus rare tumors. Common tumors are those seen with a frequency greater than 1 % in historical control data. To be considered a significant increase a p-value of <0.01 must be reached for pairwise comparison and a p-value of <0.005 for the trend test. Rare tumors are those seen with a frequency of less than <1 % in historical control data. To be considered a significant increase a p-value of <0.05 must be reached for pairwise comparison and <0.025 for the trend test.

The outcome of carcinogenicity studies is rarely an approvability issue. Positive results are included in the drug label under the section labeled *Carcinogenesis, mutagenesis, impairment of fertility*. This section describes the types of tumors seen and exposure margins relative to clinical exposures. If there is a serious concern for potential carcinogenicity, the results can be described in a "black box warning" at the top of the label. Some tumors are considered to have equivocal or limited relevance for human risk evaluation. For example, hepatic and/or thyroid tumors can be caused by drug-induced of enzyme induction. Leydig cell tumors can result from drug-induced increases in luteinizing hormone, Dopamine antagonist blockers can result in prolactin-mediated tumors of mammary, pituitary, and endocrine pancreas. These types of tumors are still included in the label but often with the caveat that their relevance to human risk is unknown.

6.10 Full CAC Meeting

The full CAC is comprised of all pharmacology/toxicology supervisors in the Office of New Drugs, approximately 25 individuals. The full CAC meets only on rare occasions to deal with unusual issues. For example, if the eCAC fails to agree on the outcome of a carcinogenicity study the full CAC would convene to render an opinion. On a few occasions, sponsors have disagreed with eCAC conclusions and requested to present their interpretation to the full CAC. A reviewer or supervisor from the regulating division would present the agency's viewpoint followed by the sponsor. The full CAC would discuss the issue and vote on the interpretation of the data. The sponsor is not present during this discussion on the subsequent vote.

6.11 The Path Forward

With our current understanding of the carcinogenic process and how chemicals produce tumors, there is growing awareness that a single study or group of studies is unlikely to provide a sufficiently robust data set for human risk assessment. Pharmaceutical chemicals provide a particularly difficult challenge as they are designed to produce a pharmacological effect. As such, simple in vitro or in vivo test systems are likely to show effects produced by these agents. The challenge remains understanding the significance of these findings for human risk.

Some attempts have been made to supplant or augment animal testing with in silico predictive tools using either statistical quantitative structure-activity relationships (QSAR approaches) or rule-based expert systems such as Multicase [11]. While remaining of potential interest at least in early screening, the training datasets for these tools remain insufficiently comprehensive and robust to offer sufficient predictive value at the present time. Another approach that is very early in its development but that shows intriguing promise is a transcriptomic approach to identification of signals of concern for carcinogenicity [51]. As more experience is gained with techniques like this with powerful computational support, additional important information may be offered that can be used to detect compounds that warrant further study.

Over the last 10–15 years, there has been an increasing awareness of the need for a global evaluation of all of the available data to enable an appropriate assessment of human risk of cancer from chemical exposure and several alternative integrative approaches have been proposed [9, 12, 22, 34]. As will be discussed in subsequent chapters of this book, a critical component of this assessment is a determination of the potential of the chemical to cause genetic toxicity. The assessment of toxic potential of new pharmaceutical agents involves a series of studies in vitro and in vivo designed to understand how to test the therapeutic efficacy of novel compounds safely in humans. In addition to the in vitro and in vivo genetic toxicity studies (discussed elsewhere in this book), a battery of studies in rodent and non-rodent species up to 6 months in rodents and 9–12 months in non-rodents generate a large body of data that can be incorporated into this global weight-of-evidence approach to assessment of carcinogenic potential. There is growing awareness that important information on pre-neoplastic events can be obtained from these studies and that data from the 2 year bioassays may not be as critical as originally thought for assessment of human risk of cancer [26].

If one examines the known human carcinogenic pharmaceuticals, it can be seen that these chemicals are either genetic toxins, immunosuppressants, or endocrine stimulants (with the exception of topically applied arsenicals) [43]. It has been suggested that histologic data from chronic toxicity studies (both in rodents and non-rodents) along with data that will inform these pharmacodynamic activities (in vitro and genetic toxicity data, in vitro and in vivo pharmacological data) can provide sufficient information to reliably predict human carcinogenic risk. A cross-industry retrospective study evaluated the predictive capability of data from chronic toxicity studies and, using this information, made a proposal for a tiered approach employing 2 year rodent studies only in those cases where decisions could not be made from the available data or questions still persisted [47]. This proposal has formed the basis for an ongoing prospective study that will inform discussions on the current ICH S1 guidance on rodent carcinogenicity assessment [44]. In this effort, companies will submit voluntary dossiers on the carcinogenic potential of their novel molecules ahead of the conduct of the 2 year studies predicting the expected outcome of these studies and assessing whether a 2 year study would alter the human carcinogenic risk assessment. If the data from this study support the conclusions from the retrospective evaluation, it may enable an alteration of the current global regulatory guidelines requiring a standard rodent bioassay for each new chemical entity. The proposal under consideration would restrict the rodent bioassay for those cases where it was not possible to obtain definitive data from in vitro and chronic in vivo assays on the carcinogenic potential of new pharmaceutical agents.

The rodent bioassay was conceived and implemented at a time when comparatively very little was understood about the carcinogenic process or how chemicals might produce human cancer. At the time of the inception of this approach, this bioassay represented the best thinking on how to address the critical challenge posed by the understanding that chemicals do, indeed, possess the potential to cause cancer and it was important to address this risk before widespread exposure to novel agents. The 60 years since this assay was first proposed and used have seen an evolution of thinking based on important advances in the sciences. The modifications to how we address this important challenge that are coming into practice and are under consideration would seem to be an appropriate direction and offer the promise of improving our ability to predict human cancer risks with streamlined processes that reflect our growing knowledge in this challenging area.

References

1. Ames BN, Gold LS (1990) Chemical carcinogenesis: too many rodent carcinogens. Proc Natl Acad Sci 87:7772–7776
2. Ando K, Saitoh A, Hino O, Takahashi R, Kimura M, Katsuki M (1992) Chemically-induced forestomach papillomas in transgenic mice carry mutant Human c-Ha-ras transgenes. Cancer Res 52:978–982
3. Boorman GA, Maronpot RR, Eustis SL (1994) Rodent carcinogenicity bioassay: past, present, and future. Toxicol Pathol 22:105–111
4. Brunner GHG, Lamberts R, Creutzfeldt W (1990) Efficacy and safety of omeprazole in the long-term treatment of peptic ulcers and reflux oesophagitis resistant to ranitidine. Digestion 47:64–68
5. Cohen SM, Cano M, Earl RA, Carson SD, Garland EM (1991) A proposed role for silicates and protein in the proliferative effects of saccharin on the male rat urothelium. Carcinogenesis 12:1551–1555
6. Cohen SM, Ellwein LB (1992) Risk assessment based on high-dose animal exposure experiments. Chem Res Toxicol 5:742–748
7. Cohen SM, Robinson D, MacDonald JS (2001) Alternative models for carcinogenicity testing. Toxicol Sci 64:14–19
8. Cohen SM (2001) Alternative models for carcinogenicity testing: weight of evidence evaluations across models. Toxicol Pathol 29:183–190
9. Cohen SM (2004) Human carcinogenic risk evaluation: an alternative approach to the two-year rodent bioassay. Toxicol Sci 80:225–229
10. Contrera JF, Jacobs AC, DeGeorge JJ (1997) Carcinogenicity testing and the evaluation of regulatory requirements for pharmaceuticals. Regul Toxicol Pharmacol 25:130–145
11. Contrera JF, Kruhlak NL, Matthews EJ, Benz RD (2007) Comparison of MC4PC and MDL-QSAR rodent carcinogenicity predictions and the enhancement of predictive performance by combining QSAR models. Regul Toxicol Pharmacol 49:172–182
12. Daktorova TY, Pauwels M, Vinken M, Vanhaecke T, Rogiers V (2012) Opportunities for an alternative integrating testing strategy for carcinogen hazard assessment. Crit Rev Toxicol 42:91–106

13. Dietrich DR, Swenberg JA (1991) Alpha 2μ globulin is necessary for d-limonene production of male rat kidney tumors. Cancer Res 16:3512–3521
14. DeVries A, Van Oostrom CTM, Hofhuis FMA, Dortant PM, Berg RJW, De Gruijl FR, Wester PW, Van Kreijl CF, Capel PJA, VanSteeg H, Verbeek S (1995) Increased susceptibility to ultraviolet-B and carcinogens of mice lacking the DNA excision repair gene XPA. Nature 377:169–173
15. Donehower LA, Harvey M, Slagel BL, McArthur MJ, Montgomery CA Jr, Butel JS, Bradley A (1992) Mice deficient or p53 are developmentally normal but susceptible to spontaneous tumors. Nature 356:215–221
16. Eastin WC, Mennear JH, Tennant RW, Stoll RE, Branstetter DG, Bucher JR, McCullough B, Binder RL, Spalding JW, Mahler JF (2001) Tg.AC genetically altered mouse: assay working group overview of available data. Toxicol Pathol 29:60–80
17. Ennever FK, Lave LB (2003) Implications of the lack of accuracy of the lifetime rodent bioassay for predicting human carcinogenicity. Regul Toxicol Pharmacol 38:52–57
18. Fung VA, Barrett JC, Huff J (1995) The carcinogenesis bioassay in perspective: application in identifying human cancer hazards. Environ Health Perspect 103:680–683
19. Goodman JI, Ward JM, Popp JA, Klaunig JE, Fox TR (1991) Mouse liver carcinogenesis: mechanisms and relevance. Fundam Appl Toxicol 17:651–665
20. Haseman JK, Seilkop SK (1992) An examination of the association between maximum-tolerated dose and carcinogenicity in 326 long-term studies in rats and mice. Fundam Appl Toxicol 19:207–213
21. Havu N (1986) Enterochromaffin-like cell carcinoids of gastric mucosa in rats after lifelong inhibition of gastric secretion. Digestion 35:42–55
22. Jacobs A, Jacobson-Kram D (2004) Human carcinogenic risk evaluation, part III: assessing cancer hazard and risk in human development. Toxicol Sci 81:260–262
23. Jacobs A (2005) Prediction of 2-year carcinogenicity study results for pharmaceutical products: how are we doing? Toxicol Sci 88:18–23
24. Jacobs AC, Hatfield KP (2012) History of chronic toxicity and animal carcinogenicity studies for pharmaceuticals. Vet Pathol 50:324–333
25. Jacobson-Kram D (2004) Use of transgenic mice in carcinogenicity hazard assessment. Toxicol Pathol 32:49–52
26. Jacobson-Kram D (2009) Cancer risk assessment approaches at the FDA/CDER: is the era of the 2-year bioassay drawing to a close? Toxicol Pathol 38:169–170
27. Klaunig JE, Kamendulis LM (2008) Chemical carcinogenesis. In: Klaassen K (ed) Cassarett and Doull's toxicology, the basic science of poisons, 7th edn. McGraw-Hill, New York, pp 329–379
28. Larsson H, Carlsson E, Hakanson R, Hattsson H, Hilsson G, Seensalu R, Wallmark B, Sundler F (1988) Time-course of development and reversal of gastric endocrine cell hyperplasia after inhibition of acid secretion. Studies with omeprazole and ranitidine in intact and antrectomized rats. Gastroenterology 95:1477–1486
29. Leder A, Kuo A, Cardiff R, Sinn E, Leder P (1990) v-Ha-ras transgene abrogates the initiation step in mouse skin tumorigenesis: effects of Phorbol esters and retinoic acid. Proc Natl Acad Sci 87:9178–9182
30. Lehman-McKeeman LD, Caudill D, Takigiku R, Schneider RE, Young JA (1989) D-Limonene-induced male rat-specific nephrotoxicity: evaluation of the association between d-limonene and α2μ globulin. Toxicol Appl Pharmacol 99:250–259
31. Lossius K, Nesheim BI (1976) Response to isoprenaline in the human pregnant and non-pregnant myometrium. Acta Pharmacol Toxicol 39:198–208
32. MacDonald JS, Lankas GR, Morrissey RE (1994) Toxicokinetic and mechanistic considerations in the interpretation of the rodent bioassay. Toxicol Pathol 22:124–140
33. MacDonald JS, Scribner HS (1999) The maximum tolerated dose and secondary mechanisms of carcinogenesis. In: Kitchin K (ed) Carcinogenicity: testing, predicting, and interpreting chemical effects. Marcel Dekker, New York
34. MacDonald JS (2004) Human carcinogenic risk evaluation, part IV: assessment of human risk of cancer from chemical exposure using a global weight-of-evidence approach. Toxicol Sci 82:3–6
35. Maronpot RR, Yoshizawa K, Nyska A, Harada T, Flake G, Jueller G, Singh B, Ward JM (2010) Hepatic enzyme induction: histopathology. Toxicol Pathol 38:776–795

36. McConnell EE, Solleveld HA, Swenberg JA, Boorman GA (1986) Guidelines for combining neoplasms for evaluation of rodent carcinogenesis studies. J Natl Cancer Inst 76:283–289
37. McConnell EE (1989) The maximum tolerated dose. The debate. J Am Coll Toxicol 8: 1115–1120
38. McClain RM (1994) Mechanistic considerations in the regulation and classification of chemical carcinogens. In: Mackey M, Hellje J, Kotsonis FN (eds) Nutritional toxicology. Raven Press, New York, pp 273–303
39. McClain RM (1995) Phenobarbital mouse liver tumors: implications of hepatic tumor promotion for cancer risk assessment. Prog Clin Biol Res 391:325–326
40. Miller EC (1978) Some current perspectives on chemical carcinogenesis in humans and animals. Cancer Res 38:1479–1496
41. Monro A (1993) How useful are chronic (life-span) toxicology studies in rodents in identifying pharmaceuticals that pose a carcinogenic risk to humans? Adverse Drug React Toxicol Rev 12:5–34
42. Monro AM (1996) Testing for carcinogenic potential. In: D'Arcy PF, Harron DWG (eds) Proceedings of the third international conference on harmonization, November, 1995, Yokohama, Japan. Queen's University Press, Belfast, pp 261–268
43. Monro A, MacDonald JS (1998) Evaluation of carcinogenic potential of pharmaceuticals; opportunities arising from the international conference on harmonization. Curr Opin Drug Saf 18:309–319
44. Morton D, Sistare FD, Nambiar PR, Turner OC, Radi Z, Bower N (2014) Regulatory forum commentary: alternative mouse models for future cancer risk assessment. Toxicol Pathol 42:799–806
45. Office of Science and Technology Policy (OSTP) (1984) Chemical carcinogens: notice of review of the science and its associated principles. Fed Regist 49:21594–21661
46. Robinson DE, MacDonald JS (2001) Background and framework for ILSI's collaborative evaluation program on alternative models for carcinogenicity assessment. Toxicol Pathol 29:13–21
47. Sistare FD, Morton D, Alden C, Christensen J, Keller D, Jonghe SD, Storer RD, Reddy MV, Kraynak A, Trela B, Bienvenu JG, Bjurstrom S, Bosmans V, Brewster D, Colman K, Dominick M, Evans J, Hailey JR, Kinter L, Liu M, Mahrt C, Marien D, Muyer J, Perry R, Potenta D, RFoth A, Sherrat P, Singer T, Slim R, Soper K, Fransson-Steen R, Stoltz J, Turner O, Turnquist S, van Heerden M, Woicke J, DeGeorge J (2011) An analysis of pharmaceutical experience with decades of rat carcinogenicity testing: support for a proposal to modify current regulatory guidelines. Toxicol Pathol 39:716–744
48. Sontag JM, Page NP, Saffiotti U (1976) Guidelines for carcinogen bioassay in small rodents. Carcinogenesis Program Division of Cancer Cause and Prevention/National Cancer Institute/ DHEW Publication (NIH), Bethesda, pp 76–801
49. Storer RD, French JE, Haseman J, Hajian G, LeGrand EK, Long GG, Mixson LA, Ochoa R, Sagartz JE, Soper KA (2001) p53 +/– hemizygous knockout mouse: overview of available data. Toxicol Pathol 29:30–50
50. Tennant RW, French JE, Spalding JW (1995) Identifying chemical carcinogens and assessing potential risk in short-term bioassays using transgenic mouse models, environ. Health Perspect 103:942–950
51. Thomas RS, Bao W, Chu TM, Bessarabova M, Nikolskaya T, Nikolsky Y, Andersen ME, Wolfinger RD (2009) Use of short-term transcriptional profiles to assess the long-term cancer-related safety of environmental and industrial chemicals. Toxicol Sci 112:311–321
52. Usui T, Mutai M, Hisada S, Takoaka M, Soper KA, McCullough B, Alden C (2001) CB6F1-rasH2 mouse: overview of available data. Toxicol Pathol 29:90–108
53. van Oosterhout JPJ, van der Laan JW, de Waal EJ, Olejniczak K, Hilgenfeld M, Schmidt V, Bass R (1997) The utility of two rodent species in carcinogenic risk assessment of pharmaceuticals in Europe. Regul Toxicol Pharmacol 25:6–17
54. Weisburger EK (1983) History of the bioassay program of the National Cancer Institute. Prog Exp Tumor Res 26:187–201

Chapter 7
Carcinogenicity Testing Strategies for Small Molecules

Abigail Jacobs and Frank D. Sistare

Abstract This chapter provides an overview of the current state of carcinogenicity testing strategies used to support marketing approvals of human small molecule pharmaceuticals. Testing strategies for biologic molecules is beyond the scope and the reader is referred to Chap. 8 by Dempster et al. In this chapter a brief history of pharmaceutical carcinogenicity testing is summarized that describes the path of evolution to our current state. The current state of pharmaceutical carcinogenicity testing strategy as defined by internationally agreed upon ICH guidelines is reviewed, including the use of transgenic mouse models in pharmaceutical carcinogenicity testing strategies. Limitations of these current testing approaches are summarized and examples are used to describe and explain the implications and impact of such limitations on practical aspects of pharmaceutical development. Often times, approaches are successfully deployed by industry scientists to support conclusions that positive rodent carcinogenicity study outcomes are related to compound class effects and are not human relevant, and examples are provided where product marketing has been enabled. Finally based on decades of such repeated experiences, a vision for a near future state pharmaceutical carcinogenicity testing strategy is described where the burdens of carcinogenicity testing may be reduced without compromising human safety, and the steps in progress to realize that vision are summarized.

Keywords Pharmaceutical • Carcinogenicity Testing • Tg.rasH2 Transgenic Mouse • ICH Guidance

---

A. Jacobs
US Food and Drug Administration, Center for Drug Evaluation and Research, Silver Spring, MD, USA

F.D. Sistare (✉)
Safety Assessment and Laboratory Animal Resources, Merck and Co., Inc., West Point, PA, USA
e-mail: frank.sistare@merck.com

M.J. Graziano, D. Jacobson-Kram (eds.), *Genotoxicity and Carcinogenicity Testing of Pharmaceuticals*, DOI 10.1007/978-3-319-22084-0_7

7.1 The Evolution of Pharmaceutical Carcinogenicity Testing

The majority of animal toxicology studies conducted to support marketing approval of human pharmaceuticals are designed to support the safe conduct of progressively longer duration clinical trials where human safety and efficacy of new drug candidates can be evaluated. Adverse findings seen only at very high exposure margins in animal toxicology studies are generally of low concern for humans. Adverse findings that are monitorable and reversible and are of questionable human significance, can most times be more definitively evaluated in clinical studies. Two types of animal toxicology studies, however, are conducted not to support clinical investigation of human safety, but rather serve as surrogates for human safety. Those are the animal studies that are conducted to assess carcinogenic potential and the animal studies conducted to assess developmental and reproductive toxicology potential. Carcinogenicity studies and developmental toxicology studies in animals are intended to reveal the likely effect expected of drug administration under relevant conditions of human use. Nevertheless, these animal studies are conducted under conditions that are designed to both pressure test and provoke evidence for potential for such toxicities at high drug exposures, as well as at relevant exposures that may more closely match human use.

Therefore, because a true assessment of human carcinogenicity potential cannot be practically evaluated across all organs and tissues, in hundreds of humans after lifetime administration of a drug under relevant conditions of use, a pragmatic approach using animals was needed to serve as a surrogate of this carcinogenicity assessment for humans, to support marketing decisions. Examples exist of pharmaceutical companies conducting 7 or 10 year cancer studies in dogs or monkeys before the period of time between 1978 and 1982 when Good Laboratory Practices [19] were established, OECD Guidelines [42] were published, and the FDA Bureau of Foods published the Red Book [16]. However, testing in mice and rats has and continues to serve as the mainstay for pharmaceutical carcinogenicity evaluation. The conditions of human use requiring rodent carcinogenicity testing; the dose and exposure of drug (and metabolites) needed to fairly evaluate a drug's human carcinogenicity potential; the duration of testing; and the species needing to be tested have each undergone evolution over the past 30+ years. Current carcinogenicity testing guidelines defining agreements reached in each of these areas were established following the launch of the International Conference on Harmonization in 1990. Negotiations for defining current regulatory carcinogenicity testing expectations for pharmaceuticals with revisions implemented through international negotiation, as supported by collective experience and data, have been described recently [50]. The three current ICH guidelines, namely S1A The Need for Long-term Rodent Carcinogenicity Studies of Pharmaceuticals [26], S1B Testing for Carcinogenicity of Pharmaceuticals [27], and S1C(R2) Dose Selection for Carcinogenicity Studies of Pharmaceuticals [28] provide recommendations on which pharmaceuticals warrant carcinogenicity testing, appropriate approaches for evaluating carcinogenicity potential, and appropriate dose selection, respectively.

The current ICHS1A guideline discusses the criteria used to determine whether an evaluation of the carcinogenic potential of a pharmaceutical is considered necessary. The guideline treats pharmaceuticals differently based on duration of exposure, establishing that for small molecule human pharmaceuticals, animal carcinogenicity studies are needed for drugs that would be used continuously, or repeatedly and intermittently for greater than 6 months. Furthermore, even for pharmaceuticals used for short durations, carcinogenicity studies may be needed when a priori concern about carcinogenic potential exists, which could include for example, chemical structure, previous compound class experience, evidence of preneoplasia in shorter term animal studies, or long-term tissue retention. ICHS1A discusses clinical duration and exposure, causes for concern, genotoxicity, route of exposure and extent of systemic exposure, and endogenous peptide and proteins and analogs. In addition ICHM3R2 [29] clarifies when in drug development that the studies should be conducted, i.e., generally to support marketing, and rarely to support clinical trials. It notes that for pharmaceuticals developed to treat certain serious diseases for adults or pediatric patients, carcinogenicity testing, if recommended, can be concluded post-approval. Some parts of ICHS1A, such as discussions of photocarcinogenicity, have been superseded by ICHM3R2 and ICHS10 [32], which no longer recommend such studies.

The current S1B guideline discusses the experimental approaches intended to assess carcinogenic potential of a pharmaceutical when such an evaluation is indicated by the criteria discussed in S1A. The S1B guideline effectively treats pharmaceuticals equally in recommending that all drugs needing carcinogenic assessment be evaluated in a 2-year rat bioassay and a 2-year or shorter term mouse bioassay. ICH S1B establishes that two species should be studied, at least one of which should be a 2-year study. It also mentions other in vivo models, such as models of initiation-promotion in rodents or models of carcinogenesis using transgenic or neonatal rodents. In the past 18 years, initiation-promotion models have not been accepted as replacements for the second species, but the use of certain transgenic and neonatal mouse models has become accepted in the United States. This guidance opened the door for a 6-month transgenic mouse study to fill the need of the second species study, used in conjunction with a 2-year rodent study, usually the rat. ICHS1B also provides general guidance on interpretation of the carcinogenicity studies pointing out the value of additional mechanistic studies to help address relevance of the results of carcinogenicity study findings to humans. Mechanistic studies have been very useful in assessing human risk from carcinogenicity findings in rodents. Cross-species receptor incidence and density for receptor-mediated effects and off-target effects, gene expression or microRNA expression, cross-species pathway analysis studies, and other studies used currently for assessment of human relevance are not specifically mentioned in the 18-year-old ICHS1B guidance.

ICH S1CR2 considers dose setting criteria for the high dose in 2-year carcinogenicity studies to include either: (1) an MTD based on toxicity endpoints, (2) a limit dose of 1500 mg/kg/day (for compounds not exceeding a daily human dose of 500 mg/day and when exposure margins of tenfold can be achieved), (3) pharmacokinetic endpoints specifying the need to reach a 25-fold exposure multiple over clinical exposure with criteria specified for comparisons of AUC in animals and humans, (4) a dose resulting in saturation of absorption, (5) pharmacodynamic endpoints that may limit high dose selection, or (6) a maximal feasible dose. All these criteria apply to studies in transgenic mice, except for pharmacokinetic endpoints. In this regard, for transgenic mouse studies there exists a data gap relating to mutual understanding and acceptance as to what would constitute a reasonable upper exposure limit to be considered an adequate test. As a result sponsors are sometimes facing a choice of conducting a transgenic mouse study at exposures that may reach hundreds-fold human exposure margins, or conducting a conventional 2-year mouse study at a 25-fold exposure margin, since regulatory position on this is evolving and presently unclear.

Although it has always been possible for a drug developer to request a waiver from carcinogenicity studies in the United States, (may be granted e.g., for short-term use, for life-threatening indications such as advanced cancer per ICHS9 [31], when values close to human exposures cannot be achieved in rodents, or for orphan drugs), ICHS6 addendum [30] specifically discusses when a biologic product can be labeled without the conduct of carcinogenicity studies. A drug developer can develop a case, based on various data sources, as to why a carcinogenicity study may not be warranted (e.g., a risk is already identified or lack of a risk seems clear, or the rodents don't have the pharmacologic activity). It is important to note that ICH Guidance S6 set the precedent allowing that for biological pharmaceuticals the opportunity exists for sponsors to explain why carcinogenicity testing would be inappropriate and waivers have been given for conducting such testing. In 2014, 10 of 11 requests for carcinogenicity study waivers of biologics were accepted by the FDA. Among the ten waivers granted are examples in each of three categories for not conducting the rodent carcinogenicity study – risk already identified, lack of risk, or rodent model is not scientifically relevant.

Generally, as more comfort has developed over time with the interpretation and understanding of recurring patterns of test outcomes, the burden of animal carcinogenicity testing for all pharmaceuticals may be expected to continue to decline. Prior to 2008, pharmaceuticals with evidence of genotoxicity could not invoke limit doses using the 25-fold exposure guidance for top dose selection. Since the most recent revision to these ICH Guidelines [28], which removed the 25-fold exposure top dose selection as a restriction for drugs with a positive genotoxicity test result, a number of publications have emerged proposing further ICH Guidance revision supporting a future state for small molecule pharmaceuticals that is analogous to that currently in place for biological pharmaceuticals under ICHS6. Such a future state is expected to reduce the resource burden needed to conduct, analyze, and report carcinogenicity studies, as well as address some limitations and imperfections of carcinogenicity testing without compromising protection of human safety.

7.2 Numerous Limitations and Imperfections of Rodent Carcinogenicity Testing Have Become Apparent Over the Years

Generally speaking the current approach as described above relying on rodents to assess pharmaceutical carcinogenicity risk potential to humans has met society's needs. One can argue that with very few exceptions rodents respond in an appropriately sensitive manner to all known human carcinogens. Immunosuppressants are variably tumorigenic in rodents, likely dependent on a variable presence of endogenous tumor virus in test animals. While arsenic and few other human carcinogens have been cited as not being convincingly carcinogenic in animals, questions of the adequacy of animal testing in such instances have been raised, and the reasonable statement has been made that "…no human carcinogens…have been tested in animals that have been shown to be unequivocally negative [25]." However, many drug-related rodent neoplasms may not be relevant to humans, especially for non-genotoxic drugs. For this reason, the specificity of the current rodent based pharmaceutical carcinogenicity testing approach has been called into question. Numerous examples of human irrelevance based on investigative toxicology study data have been made in many publications over the past 20 years, and in submissions to regulatory authorities as well, that have been used to support marketing decisions. Some explanations that have been accepted for drug-related rat carcinogenicity findings deemed of questionable human relevance to support regulatory decisions are described below.

7.3 Drug-Induced Rodent Tumors Can Be Associated with Intended Pharmacology

Some rodent carcinogenicity findings may be categorized as relating to the on-target intended pharmacologically mediated drug action, but human relevance is questioned because of the excessive and sustained nature of the pharmacologic manipulation and downstream consequences realized during the conduct of the study that are shown to be unique to the rat. Receptor distribution and potency (binding constant) can differ markedly across species, for example. Humans may have a low incidence of a receptor that is more prevalent in rats or mice, and thus would be less susceptible to effects seen in rodents under conditions of clinical use, e.g., GLP-1 agonists and thyroid C-cell neoplasms in rodents versus humans [5]. Uterine leiomyomas in mice can be caused by dopamine receptor agonists and the resulting decrease in levels of prolactin [3]. Leiomyomas of the mesovarium in rats caused by beta-2-adrenergic agonists are not thought to be relevant to humans under conditions of use [34]. Pancreatic acinar neoplasms in rats are considered to be secondary to chronic cholecystokinin stimulation, and rats are considered to be much more sensitive to this effect than are humans [21].

Mammary neoplasms may occur in rats secondary to decreases in dopamine signaling and increases in prolactin levels [20]. Sprague Dawley rats are considered to be more sensitive to this effect than humans. However, prolactin may also be increased for some of these drugs in humans. The question regarding relevance of this mechanism to humans remains a point of controversy [22].

In this same category of on-target pharmacologic mediated rodent carcinogenicity, human pathways associated with pharmacologic effects may diverge from pathways to neoplasms in rodents accounting for the lack of human relevance. Examples are provided here where human relevance is questionable under conditions of clinical use, taking into account the increased susceptibility of rodents to mechanisms of tumorigenesis, and the therapeutic margins relative to humans. Pathways for PPAR-alpha [11] and HMG-CoA reductase inhibitors [38] in humans appear to diverge from those in rodents.

Enterochromaffin-like cell tumors in male and female rats are a consequence of intended pharmacologic actions to increase stomach pH and the constant stimulation results in hypergastrinemia [2, 4]. Forestomach neoplasms in rodents are considered to be due to prolonged exposure of the drug to the forestomach in rodents. Humans do not have a forestomach, nor do they have Harderian or Zymbal glands. Therefore nongenotoxic mechanisms driving tumors specific to these organs would not be expected to be clinically relevant to humans.

Another example are alpha-glucosidase inhibitors result in a deprivation of colonic carbohydrate absorption, triggering a series of monitorable events leading to renal neoplasms in rats that can all be prevented by supplementation with glucose [24].

In the hematopoietic system, thymic lymphomas in mice that are secondary to a murine viral infection following immunosuppression may not be specifically relevant to humans. However, humans might experience other relevant effects resulting in the formation of tumors at other tissue sites from other infectious agents when the desired pharmacology of immune suppression is achieved, and effects are not necessarily a direct undesirable effect of a modifiable structure of the drug. The extent of effects depends on viral infection and viral load and not only on drug dose. Such findings of immunosuppressants in rodent studies may generally result in a labeled class warning.

7.4 Drug-Induced Rodent Tumors Can Be Associated with Off-Target and Secondary Pharmacology

In a second category are examples of drugs that result in rodent neoplasms due to off-target pharmacologic actions that are not a primary result of interaction of drug with the intended therapeutic targets. These off-target or secondary pharmacologic actions are often shown to be of questionable human relevance because many rodent hormonal pathways and hormonal levels are easier to perturb in rodents than in humans and as a result such disturbance over prolonged periods of drug administration will result in rodent tumors. F-Cell thyroid neoplasms in male rats can be secondary to drug-related liver enzyme induction and drug-related decreases in T3 and T4 with associated increases in TSH [39]. This disturbance in the negative endocrine feedback loop results in sustained stimulation of thyroid F-cells by TSH, thyroid hyperplasia, and F-cell neoplasms in rats. However, sponsors have submitted data to regulatory authorities to support drug submissions showing that TSH was not increased in humans under conditions of clinical use. Leydig cell (interstitial cell) tumors are caused in rats by various drugs when testosterone levels are depleted and LH is increased [10, 12]. This has been shown to be prevented by testosterone supplementation in rats. Testosterone has been shown by sponsors for numerous drugs to not be depleted in humans under conditions of use.

Pathways for CAR agonists appear to diverge in humans from those in rodents [15, 51]. Epidemiology data have been generated for some drugs to support irrelevance of some rat liver neoplasms (e.g., phenobarbital [35]). Most hemangiosarcomas in mice from nongenotoxic drugs probably result from rodent specific pathways [9].

Renal neoplasms in male rats related to alpha-2-u-globulin nephropathy are concluded to be rat specific [48]. Urinary bladder neoplasms have also occurred in rats secondary to pharmacology, for example with PPAR dual alpha-gamma agonists. For some such agents the neoplastic effect appears dependent on drug induced secondary mechanisms resulting in altered urine composition, precipitation of salts of endogenous minerals, and enhanced urolithiasis irritating to the bladder wall [14], while for others [36] urinary bladder neoplasms are observed in the absence of any changes in urinary sediment or mineralization. However, urinary bladder neoplasms in rodents secondary to mineralization of drug substance in the bladder upon elimination of high doses has been seen with numerous agents and is considered to not be relevant to humans when no such crystals are seen in urine in humans [8].

Drug-induced liver neoplasms in mice and rats are usually irrelevant to humans, especially when secondary to liver toxicity, or associated with a high background rate in rodents. In general, neoplasms in rodent strains with a high background control rate, are often strain specific and not usually relevant to humans (e.g., pituitary neoplasms). In these cases rodents may be rather debilitated at the end of a 2 year study which can confound interpretation of the results. Furthermore, the chance occurrence and appearance of a drug associated increase in tumor rates must be carefully considered, and upper bounds of historical control tumor rates can be very helpful in this regard.

For non-DNA reactive drugs there is usually an exposure threshold for carcinogenicity below which there is little risk for humans. Neoplasms seen only at lethal doses are generally not considered relevant to humans but when the MTD that is exceeded results in exposures that are achieved at the human recommended dose the results may not be easily dismissed. Neoplasms seen at >25× the human exposure are generally not considered relevant to humans.

Many disease states are associated with alterations (increases or decreases) in normal physiology (e.g., continued immune activation, hyperglycemia). In humans, the intent of therapeutic intervention with a drug is to bring the disease state closer to normal. However, in carcinogenicity studies, normal animals are often exposed to doses of a drug that may cause sustained changes in physiology. An example is adrenal pheochromocytomas in rats following treatment with certain SGLT2 inhibitors. These drugs cause glucose malabsorption due to off-target SGLT1 inhibition seen at the doses administered to rats, which in turn increases calcium absorption by stimulating colonic glucose fermentation and reducing intestinal pH. The resulting kidney tumors, pheochromocytomas and adrenal medullary hyperplasia seen in rats after lifetime exposure have been attributed to the sequelae of enhanced Ca++ intestinal absorption [13]. This does not happen in humans presenting initially with hyperglycemia and being given doses of SGLT-2 inhibitor which normalize blood glucose. Thus the neoplastic findings in normal rodents are not likely to be relevant to humans.

Another example is that of a drug with estrogenic activity and administered to animals with estrogen dominance in old age. Effects in animals will differ from those in humans deficient in estrogen. Uterine neoplasm development can be enhanced due to estrogen dominance in aged female rats. Somatostatin analogs, for example, can result in a high estrogen/progesterone ratio and a suppressed LH response to GnRH. This does not happen in humans.

7.5 Sponsors Are Expected to Provide Convincing Data Supporting a Conclusion That a New Test Agent Triggers the Same Key Events Critical to Driving the Same Mode of Action Previously Established Not to Be Human Relevant for Other Test Agents

It is important to note in the examples provided, that it may not be sufficient to simply point out to regulatory authorities that a rat carcinogenicity study with a new test agent yields tumors that resemble a pattern previously established not to be human relevant, such as thyroid follicular cell tumors seen in conjunction with liver hypertrophy. It is not uncommon for a new test agent discovered to cause thyroid follicular cell tumors in association with liver hypertrophy or liver tumors, but it would be important for a sponsor to show as well, that the test agent may also be a CAR activating enzyme inducer showing evidence of gene expression data, and also enhancement of thyroid hormone turnover. Data demonstrating these key events would provide confidence in the conclusion that the mode of action for the new test agent matches that of previous agents causing thyroid and liver tumors through the same key events and that the overall mode of action could be accepted and agreed to be human irrelevant. Such mode of action framework proposals have been summarized by Elcombe et al. [15] using phenobarbital as a prototypical CAR activating rodent liver carcinogen. Similar data have been described recently by Buckley et al. [6] for example for prasugrel reported to induce hepatocellular adenomas in mice that were considered secondary to enzyme induction and not relevant to human safety, and investigators [23] have proposed that liver cyp2b10 mRNA levels might be used as a biomarker of CAR activation to help address human irrelevance of rodent liver tumor findings seen with dalcetrapib. More complete investigative approaches can be seen in freedom of information dossiers for numerous pharmaceuticals shown by sponsors to be CAR inducers, including recently approved darunavir (HIV protease inhibitor) and lorcaserin (serotonin 2C receptor agonist) both with labels indicating liver and thyroid tumors seen in rats attributed to hepatic enzyme induction with limited relevance to humans.

7.6 Is a 2-Year Mouse Carcinogenicity Study Still Needed?

Most carcinogenicity assessments for pharmaceuticals that are conducted in mice are now conducted in Tg.rasH2 mice except for drugs administered by dermal application. Inadequate data exist to determine if the Tg.rasH2 mouse model is appropriate for dermal application although future studies could address this issue. However, there are very few drug products applied dermally that have resulted in skin neoplasms in the past 20 years.

It has been pointed out recently [47] that one advantage of the 2-year mouse model over the Tg.rasH2 model is that a 25× human exposure threshold can be used to set the top dose while for the Tg.rasH2 model this is not presently acceptable regulatory practice, as pointed out above. For potent drugs dosed in humans at low exposures that are very well tolerated in mice, the doses and exposures that may be needed according to current ICHS1B guidance in a 6-month Tg.rasH2 mouse study would significantly exceed the doses deemed acceptable for a 2-year mouse study. Because results in a TgrasH2 study are not lifetime exposure and in possibly more susceptible animals than nontransgenic animals, they are considered to be valuable for hazard ID and not a more precise risk assessment tool. A compilation and review of accumulated experience and data is needed, to include a comparison of the relative drug exposures necessary to drive positive tumor outcomes for the same compounds conducted in both 2-year rat and 6-month TgrasH2 studies. Such data could support a systematic data-driven approach to establishing a reasonable exposure based threshold for setting doses in TgrasH2 mice, and thereby even further reduce the occasional need to conduct a 2-year mouse carcinogenicity study.

7.7 The Expanding Role of the Tg.rasH2 Alternative Transgenic Mouse Model

Since the ICH Expert Working Group on Safety introduced ICHS1B in 1996 [27], the door was opened for scientists to choose a short or medium-term rodent study as an alternative to one of the 2-year rodent carcinogenicity studies. This guideline stimulated international collaboration to evaluate the performance and utility of newly available transgenic mouse models for carcinogenicity testing, and the results of 4 years of research with the models have been summarized in a special issue of

Table 7.1 Advantages to conducting a 6-month transgenic mouse assay

1.	Earlier insight to pharmaceutical carcinogenic potential
	Enhanced overall clinical trial safety
	Earlier re-direction of sponsors away from non-viable to more viable test candidates
	Earlier resolution of hypothetical carcinogenicity concerns
	Earlier trigger for investigative studies to understand cause for any human concern
2.	Can provide some mode-of-action understanding of positive findings
3.	May enable adoption of a strategy that eliminates the 2-year rat carcinogenicity testing timeline for clear noncarcinogens
4.	Reduction and refinements of animal use
5.	Significant savings in overall testing costs
	Reduced test article demands
	Animal husbandry costs for 6 month vs. 2 years of study activities
	Histopathology assessment costs reduced with fewer animal numbers
	Test facility space requirement demands are reduced and allow greater scheduling flexibility
6.	Reduced chance for a rodent-specific and human irrelevant false positive outcome

Reprinted from [47]

Toxicologic Pathology [49]. In those early years prior to 2003, less than 25 % of carcinogenicity study protocols being proposed in the mouse to the USFDA were requesting an alternative short or medium term mouse model, and of the models proposed in those early years only a minority were for the Tg.rasH2 model [37]. Since then the popularity of the Tg.rasH2 model in particular has grown and in 2013 and 2014, approximately 75 % of all mouse carcinogenicity studies protocols for pharmaceutical development now propose the Tg.rasH2 model [33]. Among the models evaluated, the Tg.rasH2 was deemed most versatile in its superior ability over the other new models to detect relevant human carcinogens working through both genotoxic and nongentoxic mechanisms within 6 months of dosing. The model also was shown to improve on the poor specificity of the 2-year mouse assay by avoiding detection of numerous human irrelevant rodent carcinogens [44] and this conclusion has been confirmed in a recent analysis [40] of 21 publicly available Tg.rasH2 studies used to support pharmaceutical marketing registration and of all 38 studies received by the FDA by June 2014 [33]. The scientific, strategic and business advantages to industry for conducting a 6-month transgenic mouse study rather than the standard 2-year mouse assay have been summarized in Table 7.1 (Adapted from [47]).

Initial delay in the pace of adoption of the Tg.rasH2 model appears to have been based at least partially on the perceived risk that a single spontaneous tumor appearing in a high dose animal might raise concerns regarding test compound carcinogenic potential. Since the initial roll out of the model, historical control data for spontaneous tumor incidence have accumulated [41, 43] and based on the documented relatively low incidence in spontaneous tumors, except for splenic hemangiosarcomas and alveolar bronchiolar pulmonary neoplasms, and the growing trend in the use of the Tg.rasH2 model, the industry experience with this model appears to have alleviated these concerns.

7.8 Studies in Other Species

Carcinogenicity studies are not commonly performed in hamsters. However, in one recent case in which a carcinogenicity study was conducted in hamsters, hamsters had the pharmacologic activity when rats and mice didn't. In another recent case, the hamsters had a major human metabolite not seen in rats or mice.

7.9 Future Opportunities

As described, initiatives resulting in successful modifications to carcinogenicity testing over the past 20+ years, have been driven by supporting data and shared experience between drug regulatory authorities and pharmaceutical developers through ICH negotiation. The most recent initiative launched to modify ICH Carcinogenicity Testing Guidance seeks a logical risk-based approach to eliminate the need for 2-year rat study testing of those compounds with a recognizable strong safety profile, as well as for compounds where human relevant carcinogenicity risk is expected and no benefit would be gained from the conduct of a 2 year rat study. For compounds with a strong safety profile, the approach would be based on all test evidence accumulated indicating an absence of possible off-target effects including hormonal perturbation, genotoxicity and histologic evidence from chronic toxicology studies, along with knowledge of intended on-target pharmacology. For such compounds a transgenic mouse carcinogenicity study might suffice. In this way, resources for the conduct of 2-year rat studies could be reserved for compounds with signals from chronic studies or target pharmacology indicating uncertain risk, and both transgenic mouse and full 2-year rat carcinogenicity studies should be conducted. Furthermore, potentially informative endpoints should be incorporated early and proactively to inform understanding of key events and mode of action relating to human relevance. A concept paper and business case [7] were agreed upon by EMA, FDA, PMDA, EFPIA, JPMA and PhRMA, and an ICH SI Expert Working Group was launched. The business case is based on a published proposed decision paradigm suggested by PhRMA indicating that the outcome of past positive 2-year rat carcinogenicity studies with pharmaceutical candidates could be predicted with 80 % accuracy from information available from shorter term studies [46]. This analysis followed on the heels of an earlier study of data from 80 marketed pharmaceuticals demonstrating that the absence of evidence for preneoplastic potential in all tissues in chronic rat studies was a strong negative predictor of tumor outcome in any tissue [45]. The JPMA and FDA have each conducted independent analyses of separate databases that include an additional 60 and 50 pharmaceuticals, respectively, reaching the same conclusions. This further supports the notion that the number of 2-year rat studies could be reduced under certain conditions by approximately 40 % or more, without significant risk to the public health. Each 2-year rat study: (1) uses ~600 animals (2) adds 2–3 years for completion of nonclinical studies supporting registration, and in so doing can in certain situations prolong the regulatory process and delay patient access to those new medications unless carcinogenicity studies are started at risk; (3) expends industry resources to plan, synthesize and formulate test article, conduct, analyze, report, and file (and also Regulatory Authority resources to review globally) – with a total cost of an estimated $3.75 M. The ICH S1 EWG was therefore convened, and a Regulatory Notice Document was agreed upon, drafted and posted [17] triggering a prospective test of the hypothesis. Additional supporting analyses for the current ICH initiative include an assessment of U.S. FDA drug labeling of carcinogenicity risk by Alden et al. [1] and an assessment of carcinogenicity studies for European pharmaceuticals approved for marketing between 1995 and 2009 by Friedrich and Olejniczak [18]. These authors separately concluded that carcinogenicity testing results often provided little value to the drug label that could not be otherwise obtained from an integration of shorter term study and test results.

The recently posted Regulatory Notice Document proposes that cancer risk of a new pharmaceutical can be predicted from data described above with sufficient certainty to be classified into one of three categories:

Category 1 – highly likely to be tumorigenic in humans such that a product would be labeled accordingly and 2-year rat, 2-year mouse, or transgenic mouse carcinogenicity studies would not add value.

Category 2 – the available sets of pharmacologic and toxicologic data indicate that tumorigenic potential for humans is uncertain and rodent carcinogenicity studies are likely to add value to human risk assessment. Accordingly, current S1B Guidance describes options for rodent carcinogenicity testing.

Category 3a – highly likely to be tumorigenic in rats but not in humans through prior established and well recognized mechanisms known to be rodent specific and human irrelevant, such that a 2-year rat study would not add value; or

Category 3b – highly likely not to be tumorigenic in both rats or humans such that no 2-year rat study is needed.

A prospective testing period was deemed necessary and agreed to by S1 EWG members to confirm that the same opportunities to exempt animal carcinogenicity testing are mutually visible and agreeable and accurately predictable to individual sponsors and to regulatory authorities in all three major ICH regions, before the outcomes of 2-year test results are known. Regulatory authorities will need to agree globally and practice a new process with clear criteria that involve an assessment of the adequacy and the interpretation of the data available from shorter term tests for exempting the conduct of a 2-year rat carcinogenicity study. The study data and relevant published literature expected to meet the criteria for submission of such a waiver request are described in detail as Appendix 1 in the posted Regulatory Notice Document [17] and summarized in Table 7.2. After gaining sufficient experience with processes and procedures for alignment on the new paradigm, and if outcomes are demonstrated to match predictions and expectations, then ICH members propose to adopt the new approach and modify current guidance accordingly. The accumulated actual results of approximately 2 years of

Table 7.2 Weight of evidence to be considered for a categorical assignment in the CAD

1. Knowledge of intended drug target and pathway pharmacology, secondary pharmacology, & drug target distribution in rats and humans
2. Genetic toxicology study results
3. Histopathologic evaluation of repeated dose rat toxicology studies with emphasis on chronic studies
4. Exposure margins in chronic rat toxicology studies
5. Metabolic profile
6. Evidence of hormonal perturbation
7. Immune suppression
8. Special studies and endpoints
9. Results of non-rodent chronic study
10. Transgenic mouse study (not required for CAD prediction but can contribute if available)

Adapted from Regulatory Notice Document [17]

accumulated study outcome predictions are expected to support guidance modifications by 2017/2018.

It has been pointed out recently that such a future framework involving waivers of 2-year rat studies would synergize especially well with the growing comfort in the use in particular of the Tg.rasH2 transgenic mouse model for evaluating human pharmaceutical carcinogenicity potential [40]. Waivers of 2-year rat carcinogenicity studies could lead to significant reductions in drug development timelines and shorter overall timelines to getting important new pharmaceuticals in the market to meet the medical needs of patients in those instances when the carcinogenicity evaluation can be fulfilled with early conduct of carcinogenicity studies in a single species using the 6-month Tg.rasH2 mouse. Continued use of the 2-year mouse in such circumstances would negate this advantage. In this regard, it will become important to further consider how agreement might be reached toward reasonable modification to the 25× exposure margin dose setting criteria currently limiting use of the Tg.rasH2 model for certain well tolerated pharmaceutical candidates, without introducing risk to human safety.

7.10 Conclusions

Experience collected over decades of a steadily evolving carcinogenicity testing paradigm, is steadily supporting a healthy dialog between regulatory authorities and drug sponsors as to what is necessary and sufficient to ensure human safety while being sensible with resources needed to conduct these very demanding studies. Genetically modified mice, especially the Tg.rasH2 model are becoming mainstays of pharmaceutical carcinogenicity testing, and creative investigative approaches and novel endpoints are wisely and increasingly being deployed to address questions regarding human relevance of positive rodent carcinogenicity study outcomes.

Finally, efforts have been launched recently through ICH to drive global alignment in a data driven manner toward guidance revisions that could further reduce the need for the 2-year rat study, allowing study conduct waivers when it makes sense, while maintaining the transgenic mouse for carcinogenicity testing. Before ICH guidance modifications involving waivers for the conduct of certain 2 years rat carcinogenicity studies can be considered, a prospective testing period has been launched engaging drug regulatory agencies in a world-wide collaboration with pharmaceutical sponsors to evaluate predictions of 2-year rat study outcomes on drugs in active development.

References

1. Alden CL, Lynn A, Bourdeau A, Morton D, Sistare FD, Kadambi VJ, Silverman L (2011) A critical review of the effectiveness of rodent pharmaceutical carcinogenesis testing in predicting for human risk. Vet Pathol 48:772–784
2. Ali T, Roberts DN, Tierney WM (2009) Long-term safety concerns with proton pump inhibitors. Am J Med 122:896–903
3. Alison RH, Capen CC, Prentice DE (1994) Neoplastic lesions of questionable significance to humans. Toxicol Pathol 22:179–186
4. Betton GR, Dormer CS, Wells T, Pert P, Price CA, Buckley P (1988) Gastric ECL-cell hyperplasia and carcinoids in rodents following chronic administration of H2-antagonists SK&F 93479 and oxmetidine and omeprazole. Toxicol Pathol 16:288–298
5. Boess F, Bertinetti-Lapatki C, Zoffman S, George C, Pfister T, Roth A, Lee SML, Thasler WE, Singer T, Suter L (2013) Effect of GLP1R agonists taspoglutide and liraglutide on primary thyroid C-cells from rodent and man. J Mol Endocrinol 50:325–336
6. Buckley LA, Sanbuissho A, Starling JJ, Knadler MP, Iversen PW, Jakubowski JA (2012) Nonclinical assessment of carcinogenic risk and tumor growth enhancement potential of prasugrel, a platelet-inhibiting therapeutic agent. Int J Toxicol 31:317–325
7. Business plan S1: rodent carcinogenicity studies for human pharmaceuticals. Dated and endorsed by the Steering Committee on 14 Nov 2012. http://www.ich.org/products/guidelines/safety/article/safety-guidelines.html
8. Cohen SM (1998) Urinary bladder carcinogenesis. Toxicol Pathol 26:121–127
9. Cohen SM, Storer RD, Criswell KA, Doerrer NG, Dellarco VL, Pegg DG, Wojcinski ZW, Malarkey DE, Jacobs AC, Klaunig JE, Swenberg JA, Cook JC (2009) Hemangiosarcoma in rodents: mode-of-action evaluation and human relevance. Toxicol Sci 111:4–18
10. Cook JC, Klinefelter GR, Hardisty JF, Sharper RM, Foster PM (1999) Rodent Leydig cell tumorigenesis: a review of the physiology, pathology, mechanisms and relevance to humans. Crit Rev Toxicol 29:169–261
11. Corton JC, Cunningham ML, Hummer BT, Lau C, Meek B, Peters JM, Popp JA, Rhomberg L, Seed J, Klaunig JE (2014) Mode of action framework analysis for receptor-mediated toxicity: the peroxisome proliferator-activated receptor alpha (PPARα) as a case study. Crit Rev Toxicol 44:1–49
12. Coulson M, Gibson GG, Plant N, Hammond T, Graham M (2003) Lansoprazole increases testosterone metabolism and clearance in male Sprague–Dawley rats: Implications for leydig cell carcinogenesis. Toxicol Appl Pharmacol 192:154–163
13. De Jonghe S, Proctor J, Vinken P, Feyen B, Wynant I, Marien D, Geys H, Mamidi RN, Johnson MD (2014) Carcinogenicity in rats of the SGLT2 inhibitor canagliflozin. Chem Biol Interact 224:1–12
14. Dominick MA, White MR, Sanderson TP, Van Vleet T, Cohen SM, Arnold LE, Cano M, Tannehill-Gregg S, Moehlenkamp JD, Waites CR, Schilling BE (2006) Urothelial

carcinogenesis in the urinary bladder of male rats treated with muraglitazar, a PPAR alpha/gamma agonist: evidence for urolithiasis as the inciting event in the mode of action. Toxicol Pathol 34:903–920
15. Elcombe CR, Peffer RC, Wolf DC, Bailey J, Bars R, Bell D, Cattley RC, Ferguson SS, Geter D, Goetz A, Goodman JI, Hester S, Jacobs A, Omiecinski CJ, Schoeny R, Xie W, Lake BG (2014) Mode of action and human relevance analysis for nuclear receptor-mediated liver toxicity: a case study with phenobarbital as a model constitutive androstane receptor (CAR) activator. Crit Rev Toxicol 44(1):64–82
16. FDA Redbook (1982) Latest version of 2007 at http://www.fda.gov/food/guidanceregulation/guidancedocumentsregulatoryinformation/ingredientsadditivesgraspackaging/ucm2006826.htm
17. Fed Register (2013) https://www.federalregister.gov/articles/2013/03/18/2013-06145/international-conference-on-harmonisation-proposed-change-to-rodent-carcinogenicity-testing-of Pharmaceuticals; Request for Comments
18. Friedrich A, Olejniczak K (2011) Evaluation of carcinogenicity studies of medicinal products for human use authorised via the European centralised procedure (1995–2009). Regul Toxicol Pharmacol 60(2):225–248
19. Good Laboratory Practices (2014). 21CFR 58 good laboratory practice for nonclinical laboratory studies, U.S. Code of Federal Regulations, office of the Federal Register, Govt Printing office, Washington, DC
20. Gopinath C (1995) The predictive value of pathological findings in animal toxicology studies. J Toxicol Pathol 8:89–100
21. Gould S, Scott RC (2005) 2-hydroxypropyl-b-cyclodextrin (HP-b-CD): a toxicology review. Food Chem Toxicol 43:1451–1459
22. Harvey PW (2011) Prolactin-induced mammary tumorigenesis is not a rodent specific response. Toxicol Pathol 39:1020–1022
23. Hoflack J-C, Mueller L, Fowler S et al (2012) Monitoring Cyp2b10 mRNA expression at cessation of 2-year carcinogenesis bioassay in mouse liver provides evidence for a carcinogenic mechanism devoid of human relevance: the dalcetrapib experience. Toxico Appl Pharmacol 259:355–365
24. Hollander P (1992) Safety profile of acarbose, an alpha-glucosidase inhibitor. Drugs 44(Suppl 3):47–53
25. Huff J, Chan P, Nyska A (2000) Is the human carcinogen arsenic carcinogenic to laboratory animals? Toxicol Sci 55:17–23
26. ICHS1A (1996) International conference on harmonization: the need for long-term rodent carcinogenicity studies of pharmaceuticals. http://www.fda.gov/downloads/Drugs/GuidanceComplianceRegulatoryInformation/Guidances/UCM074911.pdf
27. ICHS1B (1997) International conference on harmonization: testing for carcinogenicity of pharmaceuticals. http://www.fda.gov/downloads/Drugs/GuidanceComplianceRegulatoryInformation/Guidances/UCM074911.pdf
28. ICHS1CR2 (2008) International conference on harmonization: dose selection for carcinogenicity studies of pharmaceuticals. http://www.fda.gov/downloads/Drugs/GuidanceCompliance RegulatoryInformation/Guidances/UCM074911.pdf
29. ICHM3R2 (2010) Nonclinical safety studies for the conduct of human clinical trials and marketing authorization for pharmaceuticals. http://www.fda.gov/downloads/Drugs/Guidance ComplianceRegulatoryInformation/Guidances/UCM074911.pdf
30. ICHS6 R1 S6 Addendum (2012) Preclinical safety evaluation of biotechnology-derived pharmaceuticals. http://www.fda.gov/downloads/Drugs/GuidanceComplianceRegulatoryInformation/Guidances/UCM074911.pdf
31. ICHS9 (2010) Nonclinical evaluation for anticancer pharmaceuticals. http://www.fda.gov/downloads/Drugs/GuidanceComplianceRegulatoryInformation/Guidances/UCM074911.pdf
32. ICHS10 (2015) Photosafety evaluation of pharmaceuticals. http://www.fda.gov/downloads/Drugs/GuidanceComplianceRegulatoryInformation/Guidances/UCM074911.pdf

33. Jacobs A, Brown P (2015) Regulatory forum opinion piece: transgenic/alternative carcinogenicity assays: a retrospective review of studies submitted to CDER/FDA 1997–2014 Toxicol Pathol. 43:605–610
34. Jack D, Poynter D, Spurling NW (1983) Beta-adrenoceptor stimulants and mesovarian leiomyomas in the rat. Toxicology 27:315–320
35. La Vecchia C, Negri E (2014) A review of epidemiological data on epilepsy, phenobarbital, and risk of liver cancer. Eur J Cancer Prev 23:1–7
36. Long GG, Reynolds VL, Lopez-Martinez A, Ryan TE, White SL, Eldridge SR (2008) Urothelial carcinogenesis in the urinary bladder of rats treated with naveglitazar, a gamma-dominant PPAR alpha/gamma agonist: lack of evidence for urolithiasis as an inciting event. Toxicol Pathol 36:218–231
37. MacDonald J, French JE, Gerson RJ, Goodman J, Inoue T, Jacobs A, Kasper P, Keller D, Lavin A, Long G, McCullough B, Sistare FD, Storer R, van der Laan JW (2004) The utility of genetically modified mouse assays for identifying human carcinogens: a basic understanding and path forward. Toxicol Sci 77:188–194
38. MacDonald JS, Halleck MM (2004) The toxicology of HMG-CoA reductase inhibitors: prediction of human risk. Toxicol Pathol 32(Suppl 2):26–41
39. McClain RM (1989) The significance of hepatic microsomal enzyme induction and altered thyroid function in rats: implications for thyroid gland neoplasia. Toxicol Pathol 17:294–306
40. Morton D, Sistare FD, Nambiar PR, Turner O, Radi Z, Bower N (2014) Regulatory forum commentary: alternative mouse models for future cancer risk assessment. Toxicol Pathol 42:799–806
41. Nambiar PR, Turnquist SE, Morton D (2012) Spontaneous tumor incidence in rasH2 mice: review of internal data and published literature. Toxicol Pathol 40(4):614–623
42. OECD History. http://www.oecd.oraboutg//history/
43. Paranjpe MG, Elbekaei RH, Shah SA, Hickman M, Wenk ML, Zahalka EA (2013) Historical control data of spontaneous tumors in transgenic CByB6F1-Tg(HRAS)2Jic (Tg.rasH2) mice. Int J Toxicol 32(1):48–57
44. Pritchard JB, French JE, Davis BJ, Haseman JK (2003) The role of transgenic mouse models in carcinogen identification. Environ Health Perspect 111:444–454
45. Reddy MV, Sistare FD, Christensen JS, DeLuca JG, Wollenberg GK, DeGeorge JJ (2010) An evaluation of chronic 6- and 12-month rat toxicology studies as predictors of 2-year tumor outcome. Vet Pathol 47:614–629
46. Sistare FD, Morton D, Alden C, Christensen J, Keller D, Jonghe SD, Storer RD, Reddy MV, Kraynak A, Trela B, Bienvenu JG, Bjurström S, Bosmans V, Brewster D, Colman K, Dominick M, Evans J, Hailey JR, Kinter L, Liu M, Mahrt C, Marien D, Myer J, Perry R, Potenta D, Roth A, Sherratt P, Singer T, Slim R, Soper K, Fransson-Steen R, Stoltz J, Turner O, Turnquist S, van Heerden M, Woicke J, DeGeorge JJ (2011) An analysis of pharmaceutical experience with decades of rat carcinogenicity testing: support for a proposal to modify current regulatory guidelines. Toxicol Pathol 39(4):716–744
47. Storer RD, Sistare FD, Reddy MV, DeGeorge JJ (2010) An industry perspective on the utility of short-term carcinogenicity testing in transgenic mice in pharmaceutical development. Toxicol Pathol 38:51–61
48. Swenberg JA (1992) Alpha 2u-globulin nephropathy: review of the cellular and molecular mechanisms involved and their implications for human risk assessment. Environ Health Perspect 101(Suppl 6):39–44
49. Toxicologic Pathology (2001) 29(Supp issue)
50. van der Laan JW, DeGeorge JJ, Moggs J, Sistare FD (2013) Towards more scientific relevance in carcinogenicity testing. In: van der Laan JW, DeGeorge JJ (eds) Global approach in safety testing: ICH guidelines explained. Springer, New York
51. Yamada T, Okuda Y, Kushida M, Sumida K, Takeuchi H, Nagahori H, Fukuda T, Lake BG, Cohen SM, Kawamura S (2014) Human hepatocytes support the hypertrophic but not the hyperplastic response to the murine nongenotoxic hepatocarcinogen sodium phenobarbital in an in vivo study using a chimeric mouse with humanized liver. Toxicol Sci 142:137–157

Chapter 8
Carcinogenicity of Biopharmaceuticals

Maggie Dempster, Kris Siezen, Bas ter Braak, Willem van den Brink, Annette Emerenciana, Francesco Bellanti, Ruben G. Duijnhoven, Marcel Kwa, and Jan Willem van der Laan

Abstract The physicochemical and biological properties of biopharmaceuticals are, in many aspects, different from small molecule drugs. These differences must also be taken into account when evaluating the risk of carcinogenicity in humans. For example, because of their expected biological activity, growth factors or immunomodulators present an inherent risk for potentially enhancing tumor incidence in humans.

M. Dempster (✉)
Safety Assessment, GlaxoSmithKline, LLC., Philadelphia, PA, USA
e-mail: maggie.a.dempster@gsk.com

K. Siezen
Section on Pharmacological, Toxicological and Biotechnological Assessment, Medicines Evaluation Board, Utrecht, The Netherlands

B. ter Braak
Division of Toxicology, Leiden Academic Center for Drug Research, Leiden, The Netherlands

W. van den Brink • A. Emerenciana • F. Bellanti
Division of Pharmacology, Leiden Academic Center for Drug Research, Leiden, The Netherlands

R.G. Duijnhoven
Section on Pharmacovigilance, Medicines Evaluation Board, Utrecht, The Netherlands

Division of Pharmacoepidemiology and Clinical Pharmacology, Utrecht Institute for Pharmaceutical Sciences, Utrecht University, Utrecht, The Netherlands

M. Kwa
Section on Pharmacovigilance, Medicines Evaluation Board, Utrecht, The Netherlands

J.W. van der Laan
Section on Pharmacological, Toxicological and Biotechnological Assessment, Medicines Evaluation Board, Utrecht, The Netherlands

Division of Toxicology, Leiden Academic Center for Drug Research, Leiden, The Netherlands

Center for Health Protection, National Institute for Public Health and the Environment, Bilthoven, The Netherlands

M.J. Graziano, D. Jacobson-Kram (eds.), *Genotoxicity and Carcinogenicity Testing of Pharmaceuticals*, DOI 10.1007/978-3-319-22084-0_8

The present chapter reviews the background for this position of biotechnologically-derived pharmaceuticals. Growth factors can be seen as oncogenes, as these proteins will stimulate cell surface receptors related to cell proliferation. In this respect, ICH S6(R1) deviates from the common approach for carcinogenicity testing, as generally 2-year bioassay studies are not expected for these products. Also, for immunomodulators, the regulatory guidance acknowledges an inherent risk for cancer when immunosuppressive activity can be expected based on the pharmacology of the compound (e.g., impaired immune surveillance).

In this chapter, a few case studies are presented, illustrating different approaches in evaluating the carcinogenic potential of biopharmaceuticals. Furthermore, approaches to the translation of these findings to the human situation are discussed. Insulin-like growth factor and insulin are different in mitogenic and metabolic activity by stimulation of IGF1- and Insulin receptor A or B, respectively. Insulin analogues such as Insulin AspB10 and insulin glargine have been analyzed in this respect by novel in vitro and in vivo strategies, and this approach reveals its usefulness from a regulatory point of view.

GLP1-agonists induce thyroid C-cell tumors by a direct action at the C-cell, and we have described a pharmacodynamic/pharmacokinetic approach to model the relationship between exposure and the induction of thyroid hyperplasia or adenoma (dependent on the compound).

By virtue of their pharmacology, some monoclonal antibodies are also known to be associated with occurrence of tumors in humans, and an overview of these reported cases is also included in this chapter. The concerns of an increased cancer risk associated with medicines may arise at any time during a drug's life cycle: in early phases during development, or after many years of use in clinical practice. Pharmacovigilance represents the science and activities related to the detection, assessment, understanding and prevention of adverse effects or other drug-related problems. In this section, we review a series of medicines for which cancer has been a suspected or actual risk detected, as well as the problems that are encountered in studying and communicating such cancer risks or the uncertainties about these risks.

Keywords Carcinogenicity evaluation • Biotechnology-derived pharmaceuticals • Non-clinical evaluation • Insulin • GLP-1 agonists • Immunomodulators

8.1 Introduction

While new pharmaceuticals are constantly being developed to alleviate disease in humans, it is recognized that these pharmaceuticals can also be associated with adverse effects due to their chemical structure and/or inherent pharmacologic activity. Therefore, non-clinical and clinical evaluation of human pharmaceuticals is required to demonstrate both efficacy and safety in the intended clinical therapeutic situation. Carcinogenicity is one of the major safety concerns that is largely de-risked in the non-clinical studies. In this chapter, we focus on the carcinogenic risk of proteins produced by recombinant biotechnology, generally called biopharmaceuticals, as specific aspects make them different from conventional small molecules.

8.1.1 Carcinogenicity Risk Assessment in General: Initiators and Promoters

Experimental models in the past have helped to differentiate classes of compounds as being either tumor initiators (often detected as causing cancer by a single or at least short-term administration) or tumor promoters (usually given continuously and chronically over a long period). Initiators became known as compounds usually interacting directly with DNA, damaging its sequence, i.e. toxic for the gene. Promoters appear to need a proliferation step in the cell cycling, either by direct stimulation of proliferation, or by indirect cell multiplication e.g. as a repair of damaged tissue after a toxic phenomenon (irritation or other types of cell damage).

In the risk assessment for potential carcinogenicity of human pharmaceuticals genotoxic and non-genotoxic compounds can be differentiated. Regulators assume that genotoxic compounds have an inherent carcinogenic potential and accept these compounds as human pharmaceuticals only if their benefit outweighs their risk [1]. This is the case for some cytostatic anticancer drugs and a few anti-HIV antiviral drugs [2]. Most genotoxic compounds will therefore not be further developed as human pharmaceuticals. The field of molecular oncology is rapidly advancing and it is expected that new insights based upon this increasing knowledge will contribute to improved carcinogenic assessments of biopharmaceuticals as well as small molecules [3].

It is now well accepted that the process of cancer development is usually not a one-hit process and that the development of tumors likely requires multiple steps. Exceptions might be the hereditary retinoblastoma, also called von Hippel-Lindau disease, which has been described as a two-hit model [4], and glioneuroblastoma based on a single change in Wilms tumor1 gene [5]. Other research revealed the differentiation between "gatekeeper" genes and "caretaker" genes. Gatekeeper genes are characterized by their control of net cellular proliferation, whereas caretaker genes are involved in maintaining the integrity of the genome, and consist of repair-genes [6]. Predisposition to cancer might be related to inherited mutation of these repair genes, e.g. in Xeroderma Pigmentosa patients. Human disease characteristics are seen also in animal models, e.g. genetically modified mouse models [7, 8].

Other types of genes are called oncogenes where genetic alterations would lead to an increase in protein function and activity, and suppressor genes, where the loss of functionality is the crux, with an important point that both alleles need to be affected (loss of heterozygosity).

Examples of oncogenes are H-ras and K-ras genes. One of the existing transgenic models for standard carcinogenicity testing mentioned in the ICH guidelines is the TgRasH2 mouse, tested at the 1990s ILSI-HESI initiative Alternatives to Carcinogenicity Testing [ILSI-HESI ACT] [9].

For suppressor genes, the presence of just one intact gene might be important for resistance against cancer. An example for a suppressor gene is the p53-gene. A p53-heterozygous mouse model was established in the 1990s [10], and has been tested extensively, but was found mainly sensitive to genotoxic agents in the ILSI-HESI ACT and the EPA program.

There is now a growing insight in the types of damage that might lead to human cancer as is clear from the paper on Hallmarks on Cancer [11]. These authors have given an updated overview of all (at least a high number) of processes that are involved in the induction and progress in cancer.

Important elements among these components are the processes indicating growth and proliferation, self-sufficiency in growth signals and limitless replicative potential. In addition, insensitivity to anti-growth signals and evading apoptosis can be seen as growth-stimulating factors. These components explain on the one hand that proliferation is an important factor as a non-genotoxic phenomenon, but is on its own insufficient to lead to cancer. It should be emphasized that human epithelial cancers do not always follow a predictable histopathological sequential pattern from hyperplasia to adenoma and then to carcinoma. While this sequence is not uncommon, differences might depend on the number of spontaneous mutations.

This also clear from our large database work, that compounds with a similar pharmacological action e.g. β_2-agonists in some cases will induce benign adenomas, whereas in other cases only hyperplasia was observed (Van der Laan et al., manuscript in preparation). A Vitamin D analogue induced cell proliferation in adrenals after 6 months of administration, while pheochromocytomas were observed after 57 weeks [12].

Non-genotoxic compounds commonly enhance proliferation, either by direct receptor stimulation or by enhancing the release of proliferating factors. Even indirect stimulation of cell growth as a compensatory mechanism for cell damage can be seen as a non-genotoxic mechanism leading to cancer, e.g. by damaging bladder mucosa [13].

The methodology of modern Next Generation Sequencing allows us to assess mutations in tumor tissue, and compare the pattern of mutations among tumors from the same organ in different animals which helps explain the specificity of the mutations. Bronchud [3, 14] showed that increased number of mutations correlates with the increasing premalignant changes.

Non-genotoxic compounds might act by proliferation. Vogelstein et al. [15] recently described that spontaneous mutations occur in a variety of places, and some mutations occur in "driver-genes", whereas other mutations (the vast majority) in "passenger genes", with no direct result on the tumor-character of a cell. The number of mutations is also dependent on the age of the individual as well as the organ.

What we learned from a recent study with insulin AspB10 and IGF1, is that non-genotoxic compounds may accelerate tumor formation (in a transgenic breast cancer model in mice), and may stimulate a specific pattern of what has been called before "spontaneous mutations". See Sect. 8.3 Case studies, Insulins and IGF-1.

8.1.2 Growth Factors and Other Biopharmaceuticals

Although proliferation of cell growth cannot be seen as the single cause for cancer, its impact is high in the list of causes as explained by Hanahan and Weinberg [11]. This is recognized in the risk assessment of growth stimulating factors as medicines.

Growth factors can be seen as a category of oncogenes, as these proteins stimulate cell surface receptors [for a review see Pan and Godwin] [16] leading to signal transduction relating to cell proliferation. Classically the definition of an oncogene is a gene that will transform the cell with some attributes of malignancy. With insight that is more recent we now know that the effect of proliferation by a growth factor depends on the cellular context. Growth factors and extracellular mitogenic signals are identified as Platelet-Derived Growth Factors (e.g. PDGFβ), Fibroblast Growth Factors (e.g. FGF-3/INT-2, FGF4/HST), WNT (e.g. WNT-1, WNT-2), Epidermal growth factor (e.g. EGF, TGF-a) or cytokines (e.g. Interleukin-2, Granulocyte-Macrophage-Colony Stimulating Factor). Another class of oncogenes can be identified as cell surface receptor, such as the EGF receptor family (EGFR, ERBB2 [HER-2/neu]), PDGF receptor family, VEGF receptor family, but also the insulin-receptor family. These receptor families are receptor tyrosine kinases. The receptor domain is located extracellularly and binds the growth factors, whereas the kinase domain is located intracellularly. This kinase domain is now a target for numerous modern anticancer agents, known as more- or less specific tyrosine kinase inhibitors.

Growth factors are polypeptides stimulating cell surface receptors very specifically with high affinity. Unlike endocrine hormones the specific growth factors usually have a local target, autocrine, paracrine or juxtacrine in character. Keratinocyte growth factor (KGF) is such a paracrine growth factor. Kepivance® is on the market as recombinant KGF, and has been specifically evaluated with epithelial cell lines and human carcinoma xenografts. While the results of these nonclinical studies confirmed a potential tumorigenic risk, they also provided important insight that there was a low likelihood that this would occur in humans [see below for further details] [17].

It is important to keep in mind that with systemic administration there might be barriers for growth factors to reach their targets, e.g. the extracellular matrix. Sometimes matrix components are actively involved in the interaction between growth factor and receptor. For example, heparin is involved in the interaction between fibroblast growth factor (FGF) and FGF-receptors [18].

These aspects of growth factors and their receptors are what create inherent risks for a carcinogenic potential.

8.2 Regulatory Guidance

The ICH S6(R1) guideline [19] is the primary source of advice for biopharmaceuticals and provides recommendations for the types of non-clinical studies with which to evaluate potential for toxicity. Because of their unique biological and physiochemical characteristics, ICH S6 recommends a scientifically based case-by-case approach. As discussed in detail above, although biopharmaceuticals are not genotoxic and therefore not expected to be ‘complete carcinogens’, chronic administration could potentially result in an increased risk of tumor promotion and/or growth based on their expected pharmacologic activity [20]. Evaluation of the carcinogenic potential for any new chemical entity depends on both its intended clinical duration of use, type of disease and specific concerns based on its pharmacological properties including genotoxicity as recommended in ICH S1A [1]. Although ICH S1A primarily addresses small molecular weight compounds, several scenarios are presented when a rodent 2-year bioassay should be considered for biopharmaceuticals. These include (1) different biological effects observed between the recombinant protein and the endogenous product; (2) structural differences between the recombinant product and natural product; and (3) recombinant products administered at pharmacologic doses greater than expected endogenous levels. These scenarios focus on recombinant proteins (e.g. growth factors, hormones and interferons) intended for replacement or augmentation therapy and do not pertain to other biopharmaceuticals such as mAbs and fusion proteins.

In the original ICH S6 guideline published in 1997, it was recognized that depending on the duration of clinical dosing, patient population and/or exaggerated pharmacology, an assessment of carcinogenic potential may need to be considered. The guidance suggests that a 2 year rodent bioassay assuming that relevant pharmacological activity can be sustained could provide useful data if the accumulated safety database is not sufficient to determine the potential for carcinogenicity. Since the pharmacology of certain classes of drugs, such as growth factors (see above) and immunosuppressive agents [19, 21] could represent a potential carcinogenic risk following chronic administration, the purpose of the S6 guideline was to offer alternatives rather than to default to the rodent bioassay to provide an appropriate carcinogenic risk assessment.

Differences in interpretation and implementation of the original ICH S6 guideline as confirmed by the increasing number of examples of opposing recommendations from the global regulatory regions led to an addendum of the guideline [19, 22]. The purpose of the addendum was to clarify several topics, including the carcinogenicity section while still maintaining the flexibility, and the case-by-case approach mandated in the original guideline. In the S6 addendum, with respect to the carcinogenicity, the section was expanded to provide more detail and to offer suggestions for different scenarios. Similar to the original guideline, a product-specific assessment of carcinogenic potential should be considered based on the duration of dosing and/or mechanism of action of the biopharmaceutical and when there is a potential concern, a number of approaches should be considered. This product-specific assessment should be based on accumulated nonclinical data and knowledge of the intended mechanism of action with the product. Literature data (from knock out animals, human genetic diseases), information from similar targets or class effects, and clinical data can provide useful information with which to base a risk assessment [19].

8.2.1 General Practical Advice

An advantage with respect to nonclinical safety strategy for biopharmaceuticals distinct from small molecular weight compounds is their type of toxicity, i.e. toxicity associated with biopharmaceuticals is primarily limited to exaggerated pharmacology and therefore potential toxicity should theoretically be easier to predict [23, 24]. Accordingly, a specific carcinogenicity risk-assessment strategy should be defined early in a development program and be updated periodically as nonclinical information accumulates. This risk assessment can be added to briefing documents submitted to Health Authorities and in addition to communicating risk, suggestions for a risk management plan which may include clinical or post-marketing monitoring and labeling proposals. A good example of a specific product carcinogenic strategy is a critical assessment from both a scientific and practical point of view to appropriately assess the potential carcinogenicity of Interleukin-10 [25]. Their assessment of the known biological activity across different species concluded that chronic administration of IL-10 would not be expected to be associated with a carcinogenic risk. In addition, a critique of the various other models such as transgenic mice and xenograft models were unlikely to provide relevant data.

Products will generally fall into one of three categories, those in which there are sufficient data for assessing potential carcinogenicity, those where there are insufficient data and those in which the mechanism of action infers a potential for carcinogenic risk.

Those products that enable an appropriate risk assessment without the need for additional nonclinical studies are those in which no data from either the repeat dose toxicity studies or alerts from a review of the literature (including knock-out animals, target biology) indicate that the candidate pharmaceutical is involved in either growth potential or cell proliferation. Recombinant proteins that are identical to the native protein sequences such as coagulation factors used for replacement therapy could be examples where additional studies may not be required. IL-10 discussed above is another example. Agents such as antagonists to growth factors, e.g. anti-VEGF mAbs meant to inhibit angiogenesis are also examples and in fact are used as an anti-cancer therapy.

Chronic administration of growth factors and immunosuppressive agents, on the other hand represent a potential concern for carcinogenic potential. For some of these types of products, i.e. growth factors, evaluation of transformed cells or xenograft models may be useful alternatives to the longer term in vivo repeat dose toxicity studies. Recombinant human keratinocyte growth factor (rHuKGF) for example was evaluated using human tumor cell lines (using both KGF+ and KGF- cell lines), a mouse xenograft model and a modified transgenic rasH2 (Tg.rasH2) model. In vitro results were not completely consistent; some of the tumor cell lines were positive and some were negative. Some of those positive cell lines were further evaluated in the xenograft model and one; possibly two of the six/seven that were positive showed a modest dose-dependent increase in growth [17, 20, 26, 27]. The rasH2 transgenic assay, however, was negative. In the repeat dose toxicity studies in rats, gastric hyperplasia and hypertrophy provided evidence of the expected pharmacologic activity of rHuKGF. Although the rasH2 transgenic model was negative, since increased proliferation was observed in at least one xenograft model, the USPI label (US packet insert) states under Warnings and Precautions "the effects of Kepivance® on stimulation of KGF receptor-expressing, non-hematopoietic tumors in patients are not known. Kepivance® has been shown to enhance the growth in human epithelial tumor cell lines in vitro and to increase the rate of tumor cell line growth in a human carcinoma xenograft model" [27]. A similar strategy was followed for recombinant human erythropoietin (rHuEPO) in which rHuEPO was incubated with various cell lines and evaluated in numerous xenograft models and in one mouse surrogate carcinogenicity study [28]. Erythropoietin produced no effect in any of these models [20]. Despite all of the negative data, the USPI carries a black box warning based on shortened overall survival in patients with certain types of cancers [29]. The Summary of Product Characteristics (SmPC) added that "erythropoietin receptors may be expressed on a variety of tumor cells" [20]. Similarly for immunosuppressive agents an increased risk of malignancy is generally accepted, which is why ICH S6(R1) recommends that with these types of compounds, the potential hazard is best addressed through appropriate product labeling and clinical risk management practices.

There will be targets, particularly novel ones, in which there are insufficient data available with which to conclude that no additional non-clinical data are needed. In this case, a more extensive evaluation may be necessary and may include the option of additional non-clinical studies which might include a 2-year rodent bioassay.

In summary, it is good practice to begin to consider the long term consequences of a particular target early in the discovery process. At this time, a review of literature, in conjunction with in vitro and in vivo efficacy data can provide knowledge of what level of potential concern exists for carcinogenic risk. For a target that has an obvious risk (i.e. growth factor agonist), additions to repeat dose toxicity studies can be included such as proliferation indicators. In addition, transformed cell lines, xenograft models, transgenic mouse models and of course the 2-year rodent bioassay can all be considered. However, these models need to be accurately characterized and scrutinized as to their relevance to the patient population. For example, questions such as whether target deficient mice which have reduced levels of, but are not depleted of target cells are a relevant model. Are there basic differences in physiology between human and rodents? The lack of cross reactivity with rodents could push for the need to use surrogate models and how representative are those models to human. Finally the lack of background data for many of the transgenic mouse models could lead to misinterpretation of findings. Therefore decisions about the type of studies need to consider the relevance of the animal data to human and that the conduct of the study should be designed to mitigate the concern or the label should reflect the concern.

8.3 Case Studies

8.3.1 Insulins and IGF-1

Insulin is a naturally occurring compound and also a known growth factor essential for normal functioning of metabolic processes. For a long time it was thought that insulin binds the insulin receptor and induces only metabolic effects, whereas IGF1R activation by IGF1 and IGF2 would induce mitogenic activities. With the discovery of the different isoforms of the insulin receptor it was found that activation of the insulin receptor A (IRA) could also induce mitogenic effects [30].

Like IGF2, IGF1 is able to activate the IGF1R but it has a low affinity for IRA and IRB. Insulin can only bind to IRA and insulin receptor (IRB) (Fig. 8.1).

While the intended pharmacological action of insulin is mediated through IRB, the mitogenic potential of insulin and insulin analogues are related to their affinity for and downstream effect via the IRA and IGF1R.

There are two distinct and well studied signaling cascades, the PI3K/Akt and Erk/MAPK. PI3K/Akt is thought to have a major role in metabolism, whereas Erk/MAPK leads to the more mitogenic effects.

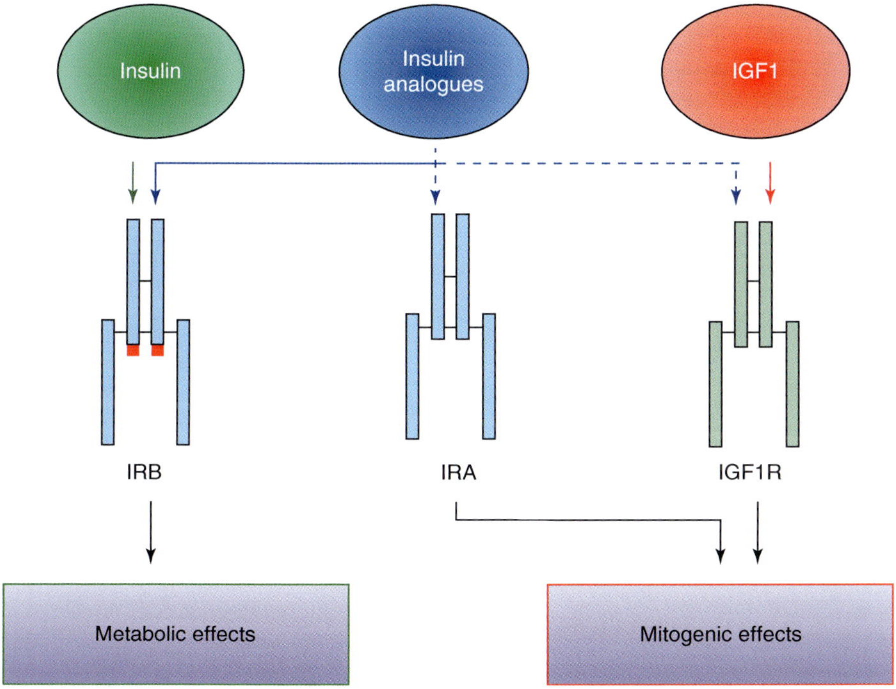

Fig. 8.1 Schematic overview of crosstalk of the insulin and IGF receptors. The *red box* in the α-subunit of the IRB represents exon 11

The phosphatidylinositol 3-kinase (PI3K) protein complex gets phosphorylated at the regulatory subunit p85α, the catalytic subunit of this complex produces PI(3,4,5)P3 (PIP3), a molecule that recruits phosphatidylinositol dependent protein kinase 1 (PDK1) to plasma membrane. PDK1 phosphorylates and activates the kinase Akt. Akt activates and inactivates a whole range of different proteins by phosphorylation including p27kip (which inhibits cell cycle inhibition), PDE3B (which induces lipolysis through PKA), FOXO1 (induces gluconeogenesis), Foxa2 (inhibits gluconeogenesis), AS160 (induces translocation of Glut-4, which induces glucose uptake), GSK3 (reduces glycogen synthesis and lipogenesis), AMPK (induces lipogenesis), TSC1, TSC2 (induces mTORC1 activity which include mitogenic effects), BAD (apoptosis), transcription factor FKHR (which induces mitogenic effects and c-Jun and JNK.

The Erk/MAPK signaling cascade is initiated by phosphorylation of IRS1/2 and Shc which recruit the SOS/Grb2 complex. This complex trigger activation of the membrane bound GTPase Ras, which in turn activates raf, which will phosphorylate Mek, which will phosphorylate Erk1 and Erk2. Finally the activated Erk will phosphorylate numerous substrates (Elk1, c-myc, SRC1, Pax6, STAT3 and c-FOS), these substrates are involved in the onset of the transcription machinery that will lead to the mitogenic effects (angiogenesis, cell proliferation, cell survival, protein synthesis and cell growth).

Insulin analogues are widely used to control the blood glucose levels in a more steady and precise manner than it would be possible with regular human insulin injections. Small variations have been incorporated in the insulin molecular structure to change its ADME (administration, distribution, metabolism and excretion) characteristics. One of the first insulin analogues developed was insulin AspB10, a fast acting insulin analogue with its histidine-B10 residue replaced by an aspartic acid residue [38]. This molecule was known to have an increased binding activity for IR, but harbored also a 7–10 times higher binding affinity for IGF1R [39]. Several studies reported an enhanced proliferative behavior of cancer cells after stimulation with AspB10 compared to regular insulin [40–44]. It is thought that this mitogenic activity of AspB10 was caused by up regulation of the p70S6K signaling pathway [45].

Also in vivo studies have been performed to study the mitogenic actions of insulin AspB10 and directly after the first in vivo study reported an increased incidence of mammary tumors in female rats; the development of AspB10 was discontinued [46]. Also some recent studies could confirm the increased carcinogenic potential of insulin AspB10 using different mouse and rat models [42, 47, 48].

This finding of mammary tumors for AspB10-insulin also drew the attention of the regulatory authorities for human medicines, and they prepared a position paper to deal with this issue [49]. In fact, the Position paper emphasized that sponsors should scientifically support their approach to evaluate the carcinogenic potential of new insulin analogues, without choosing by default to conduct a 2-year rat study. A stepwise approach starting with in vitro receptor pharmacology is proposed, and the approach chosen by Ter Braak et al. [43, 48] reflects this.

Since AspB10, many other insulin analogues have been developed and all of them have been thoroughly tested for possible carcinogenic side effects.

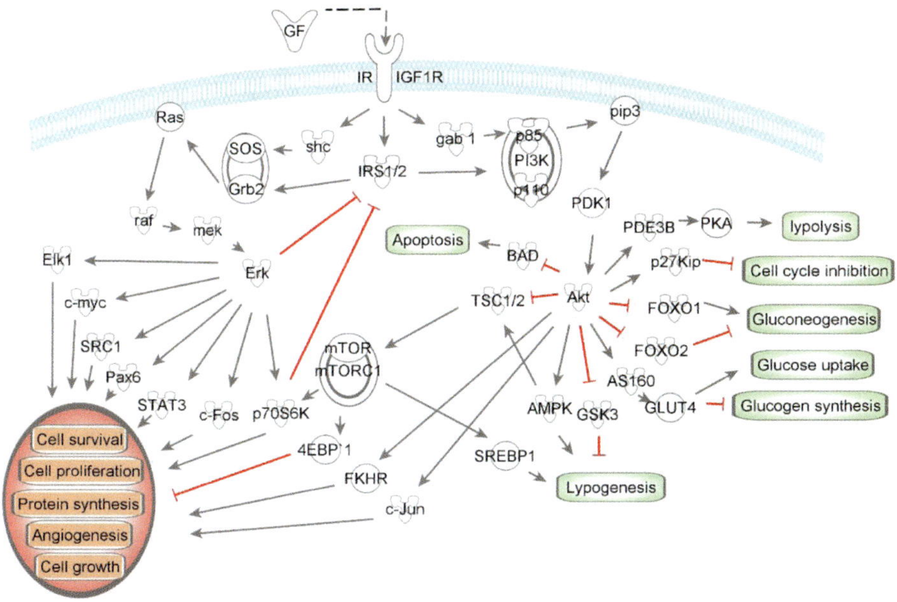

Fig. 8.2 Insulin/IGF signaling network. Ligands bind to the IR/IGF1Rs which induces phosphorylation of the downstream effector molecules inducing activation of canonical signaling pathways like PI3K-Akt and Erk-MAPK, which eventually can induce a whole range of mitogenic or metabolic effects [31–37]

One insulin analogue (insulin Glargine) has obtained extra attention since, like insulin AspB10, this compound has an increased binding affinity towards the IGF1R [50, 51]. Some in vitro studies showed increased mitogenic effects for insulin glargine [43, 52–54], where others found no increased mitogenic potential of insulin glargine compared to regular human insulin [55, 56]. A likely explanation for these contradictory findings is that in the presence of serum, glargine is rapidly metabolized by endopeptidases into M1 and M2 (M1 after removal of the two arginines, M2 with additional deamination of threonine at position B30) which have a low mitogenic potency [57]. Therefore, depending on the experimental set-up either the mitogenic potency of glargine or M1 and M2 has been measured.

The results from in vivo studies regarding the carcinogenic potential of glargine have generally shown negative results [47, 58, 59] although the most recent study using a conditional breast cancer mouse model did reveal up regulation of mitogenic MAPK-signalling pathway similar to AspB10 and IGF1 [48].

The examples described above of insulin Asp B10 and insulin glargine, underline the importance of appropriate testing for carcinogenic potential of new insulin analogues. Currently, the focus lies on receptor binding and functional effects for the IR isoforms and IGF1R, but characterization of downstream signaling pathway activation as shown in Fig. 8.2, also seems to have important predicting potential. The transcriptomic changes downstream of these pathways can be used as a tool to predict the biological direction of cells and tissues. A microarray provides a high-throughput platform for the development of such genetic classifiers as described [60], and for drug screening purposes quantitative PCR is a cheap alternative to quickly evaluate the mitogenic potential of these growth factors. Additionally chronic in vivo experiments could be useful to evaluate the in vivo carcinogenic effects of insulin analogues. While the use of humanized cancer models might improve the accuracy of carcinogenicity assessments and reduce animal numbers, further evaluation is needed to demonstrate clinical relevance of these models. Since cancer is a heterogeneous disease it is essential that the tumors in these models are properly characterized preferably by combining different omics approaches (as was done in ter Braak et al. 2015, [NGS] (manuscript in preparation).

8.3.2 GLP1-Agonists

In the last few years, GLP-1 receptor agonists have been on the market for treatment of type II diabetes. These drugs are designed to improve the balance between insulin and glucagon secretion, to lower gastric emptying and to reduce appetite. Several products are currently on the market AstraZeneca has marketed Byetta® (exenatide fast release [FR]), and Bydureon® (exenatide slow release [SR]), Novo-Nordisk has marketed Victoza® (liraglutide) and Sanofi has marketed Lyxumia® (lixisenatide) [61–64]. Eli Lilly received a marketing authorisation for Trulicity® (dulaglutide) in the summer of 2014 [65]. In addition, Eperzan® (albiglutide) was approved in 2014 [66].

Exenatide and lixisenatide are synthetic peptides based on solid phase peptide synthesis and are not biopharmaceuticals in the sense of a recombinant biotechnologically-derived product [61, 64]. Liraglutide is a recombinant protein produced in *Saccharomyces cerevisiae* [63]. Dulaglutide is a recombinant protein, generated in *Chinese Hamster Ovary* (CHO) cells, and consists of two chains in one molecule, with IgG_4 Fc-parts as the bases [65].

Carcinogenicity studies in rats and mice showed a special effect of these GLP-1 agonists with respect to induction of tumors, i.e. an increased risk for thyroid C-cell carcinogenicity. This was seen for the first in class, exenatide, with a small risk with the fast-release form Byetta, but with an enhanced risk in the slow release form, Bydureon. It was also observed with liraglutide (Victoza), which has slower elimination than exenatide. The latter two products thus have higher chronic exposure then Byetta.

A mode of action (MOA) was identified relating GLP-1r agonist exposure to C-cell carcinogenicity (Fig. 8.3). This MOA was confirmed to be GLP-1 receptor specific. Basically, the MOA tells us that binding of the agonist to GLP-1 receptors on the surface of the C-cell leads to increased cAMP concentrations in the cell. Apparently this stimulation of the cAMP concentration leads to an increase of the mRNA for calcitonin, and the production of calcitonin itself by the C-cells.

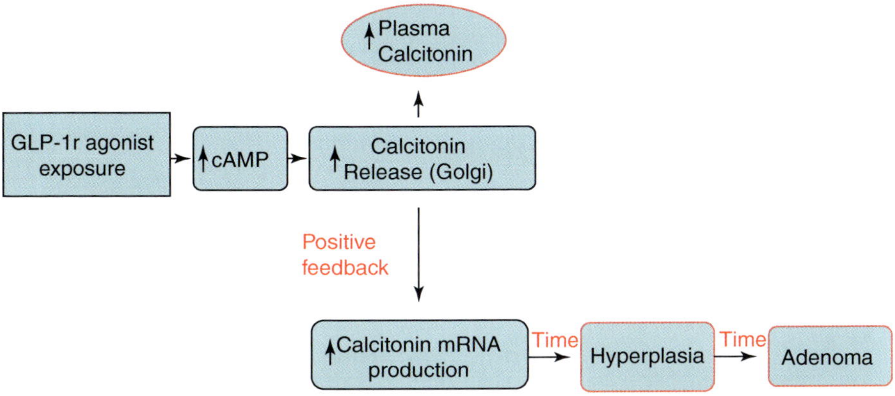

Fig. 8.3 Schematic representation of the mechanism of action relating GLP-1 receptor agonist exposure to C-cell carcinogenicity

This increase in mRNA is associated with hyperplasia of the C-cells, and eventually some cells are transformed to adenoma and carcinoma (see section 8.1.1).

The question arose as to which underlying factors determine C-cell carcinogenicity and how these factors may explain the differences among the various GLP-1 analogues. An additional question is whether the effects seen in rodent are relevant to humans. Indeed, generally a linear concentration-effect relationship is assumed when scaling the animal observations to the human situation.

However, receptor-mediated effects usually have non-linear concentration-effect sigmoid relationships. In addition, animals may have a different sensitivity that is not quantified in the standard approach of carcinogenicity assessment. A technique to answer such questions is mechanism based pharmacokinetic-pharmacodynamic (PK-PD) modeling [67–70]. We developed such a framework for assessment of GLP-1 receptor induced C-cell carcinogenicity by producing a PK model for the various GLP-1 products in animal studies, initially for exenatide FR [Byetta®] and liraglutide [Victoza®] [71]. In a following step, the liraglutide PK model was successfully extended with a PD model, describing the concentration-effect relationship with plasma calcitonin as a biomarker.

Based on this framework a PKPD model was developed for both exenatide FR and exenatide SR (Bydureon®). Given the lack of information on calcitonin levels in this case, a logistic regression model was developed linking the chronic exposure of exenatide directly to adenoma incidence.

The modeling approach provides a promising method to investigate the underlying mechanisms of the exposure response relationship in toxicological problems regarding a single drug or a drug class. The analysis conducted thus far also illustrates the importance of applying this approach from the very beginning of the development of a pharmaceutical candidate. During the application of the modeling approach some data gaps or weaknesses in the design of a study program (e.g. lack of calcitonin data) became clear. Further steps have still to be explored to apply this approach for the translation to the human situation.

GLP-1 receptors have been proven to be co-localized with thyroid C-cells in humans, but the density is much lower, although not absent. Data are available from long-term treatment with liraglutide in nonhuman primates. After 52 weeks no effects of liraglutide on C-cell proliferation was observed [63], suggesting that primate C-cells are less sensitive to proliferation induced by stimulation of GLP-1 receptors. The relevance of the rodent tumors for humans is likely to be low but cannot be completely excluded (See Sect. 8.4. Pharmacovigilance, risk management and regulatory actions taken).

8.3.3 Immunosuppressive and Immunomodulatory Agents

Immunomodulatory therapeutic monoclonal antibodies currently comprise a large portion of biopharmaceuticals available for clinical use and are widely prescribed. The theoretical risk for long term use of these agents is the risk of malignancy, in particular lymphoma which could result from a disruption of the immune system's host defense [20, 72]. Increased tumor risk, primarily lymphomas has been associated with genetic immunodeficiencies such as severe combined immunodeficiency (SCID) and Wiskott-Aldrich syndrome [72, 73]. In addition, an increased incidence of tumor types has been observed as a result of viral infections. For example, infection with HIV is associated with an increase in tumors such as Kaposi's sarcoma and Burkitt's lymphoma and in the case of HPV infection; an increased risk of cervical cancer has been observed [74, 75].

Cyclosporine, a commonly prescribed medicine for diseases such as psoriasis, has been shown to induce lymphomas in monkeys [76]. In addition, clinical use with azathioprine and cyclosporine in renal transplant patients have been associated with an increase in a number of tumor types, the majority being lymphoma and skin tumors with lower incidences in other organs such as lung, cervix, brain, etc. In one data set from the Human Kidney Transplant Registry (1971–1976), incidences of lymphomas and skin tumors could be as high as 30–40 and 4.2 times the general population, respectively [74, 77].

With respect to biopharmaceuticals, OKT3, the first therapeutic monoclonal antibody available in 1986 was used as an anti-rejection drug for organ transplantation [78, 79]. Its primary in vivo action is to opsonize the circulating lymphocytes by binding to the CD3 receptor. These cells are subsequently removed by the reticuloendothelial cells in the liver and spleen and are non-functional when they reappear [80]. Because OKT3 cross reacts only with human, chimpanzee and gorilla CD3, no animal chronic toxicity or carcinogenicity data are available, clinical experience with OKT3 provides the bulk of the safety database [78]. Shortly after the introduction of OKT3, a sharp increase in post-transplantation lymphoproliferative disorder (PTLD), a well recognized complication of immunosuppression became apparent in cardiac transplant patients [81]. The incidence of PTLD was higher in OKT3-treated patients than in patients who did not receive OKT3, 11.4 and 1.3 %, respectively. According to Swinnen's multivariate analysis [81], the only factor that was significantly associated with PTLD was the use of OKT3. A dose response relationship was also evident in that 35.7 % of patients that received a cumulative dose of more than 75 mg OKT3 had PTLD versus 6.2 % patients that received a cumulative dose less than 75 mg OKT3. In addition, the interval between OKT3 treatment and PTLD emergence was shorter, often 1–2 months in patients with the higher cumulative dose although definitive conclusions cannot be confirmed given the low number of patients [81]. Primary infection or reactivation of Epstein-Barr virus (EBV) is also thought to play a role in the pathogenesis based on its presence in tumor tissue obtained from lymphomas and the development of PTLD is thought to result from an inadequate T-cell control over EBV-driven B cell proliferation [72, 81]. Although various lymphomas are associated with EBV there are differences in disease burden. For example EBV can be detected in >90 % of cases from patients that develop lymphoma within 1 year following transplantation whereas EBV was detected in approximately 50 % of patients who developed lymphomas after 1 year following transplantation [72, 82, 83]. In addition, there was no difference in the frequency of reactivation of EBV between those patients dosed with OKT3 that exhibited PTLD and those that did not [81].

The potential association with reactivation of viruses has also been observed in animal studies. Administration of alefacept, a fusion protein which it binds to CD2 inhibits the CD2/LFA-3 interaction thus resulting in T cell depletion. In the 12 month monkey toxicity study, a lymphoma was noted in one female. In addition, B cell hyperplasia was observed in some of the other monkeys. Reactivation of lymphocryptovirus (LCV) was thought to be related to the lymphoproliferative changes as it is known that LCV infection can lead to B-cell lymphomas in immune suppressed monkeys [20, 84].

Abatacept, a fusion protein which inhibits T-cell activation by blocking the interaction the antigen presenting cell with CD28, produced an increase in the incidence of malignant lymphomas at all dose levels and mammary gland tumors in female mice at the mid and high dose levels in a 2 year bioassay. Further analysis of the mice showed that they were infected with murine leukemia and mouse mammary tumor viruses. Similar to the example discussed above, it is known that reactivation of these viruses can occur in immunosuppressed mice. However, administration of abatacept to monkeys for 12 months did not result in lymphoproliferative disease even though there was evidence of immunosuppression (depletion of germinal centers in lymph nodes and spleen) and the monkeys were known to be infected with LCV. Therefore, although there appears to be an association between immunosuppression and reactivation of latent viruses leading to an increased risk in malignancy, assessment of the relevant clinical risk are difficult due to the variability in results [20].

Host defense mouse models have also been used to evaluate immunomodulatory agents. In one published example, keliximab (anti-CD4 mAb) was tested in a B16 melanoma experimental metastasis model. Although administration of the positive control (a pan T-cell antibody) increased the number of lung metastases, keliximab, a selective CD4+ mAb had no effect [85].

In general, given the variability in results in animal studies and the lack of confidence that a negative finding in a non-clinical model can mitigate or eliminate the theoretical risk, the conduct of these animal studies is not recommended and hence few published examples exist. Instead the theoretical risk is typically outlined in the product label under Warnings and Precautions including a defined Risk Management plan. Therefore for many of the immunomodulatory agents the current clinical practice is to monitor patients post-marketing as is the case with the anti-TNF therapies.

8.3.3.1 Tumor Necrosis Factors (TNF) Inhibitors

TNF was initially isolated as a key cytokine involved with the necrosis of tumors, hence its name [86, 87]. Current understanding of the biology of TNF is that it is believed to play a regulatory role in inflammation and host defense [72, 88]. However, TNF has also been shown to exhibit diverse effects on tumor biology which are not completely understood. In contrast to inhibition of tumor growth, locally produced TNF (i.e. within the tumor microenvironment) has been shown to promote DNA aberrations resulting in maintenance of cancer growth and spread [88, 89]. In clinical trials, high levels of TNF in patients with chronic lymphocytic leukemia were associated with increased tumor spread [89, 90]. Therefore, depending on the situation, TNF can be viewed as either an anti-cancer agent or as a tumor promoter. In fact, data from several Phase I studies have shown a stabilization of disease in some of the patients with progressing advanced cancer [88, 91]. Although its exact function still needs to be clarified, anti-TNF mAbs have demonstrated effective control of auto-immune diseases such as rheumatoid arthritis [RA] and inflammatory bowel disease [IBD] [92, 93]. Because of its involvement in host defense, a number of studies have been conducted in RA patients to evaluate whether an increased risk in malignancy is associated with anti-TNF therapies. Conflicting results however have been observed (Table 8.1). The primary reason for this discrepancy could be due to the method of data analysis. Some of the investigations analyzed data using randomized clinical trials (RCTs) whereas national clinical, health, and demographic country registers were used to collect data in other studies [94–104]. Analyzed data from early RCTs in RA patients showed a dose-dependent increased risk for malignancy following the use of infliximab and adalimumab and a trend for an increase for etanercept [95, 96]. A subsequent study that evaluated a greater number of RCTs (63 versus 9), additional anti-TNF therapies to infliximab, etanercept and adalimumab (golimumab, certolizumab), and other mAbs (rituximab [anti-CD20] and tocilizumab [anti-IL6]) did not observe an increased risk of malignancy in patients treated for at least 6 months [104].

In addition, increased malignancy was not elevated following anti-TNF treatment in cohort studies using country registries [89, 97–100, 105]. In one of these studies, 6366 RA patients who had recently begun anti-TNF therapy were followed in some patients for up to 6 years for a total of 25,693 person-years and except for the first year of follow up, no differences were noted among the 3 anti-TNF drugs (adalimumab, etanercept and infliximab). During that first year, as compared with the cohort of unselected, biologics-naïve patients, patients receiving adalimumab exhibited an

Table 8.1 Selected studies to assess potential risk in rheumatoid patients

Study type	Total # patients or clinical trials	Exposure interval years	Tumors evaluated	Results	Reference
Cohort NDB USA	18,572	Median: 1.25 Range: 0.1–4.5	Lymphoma	Lymphoma ↑ in RA pts Although ↑ lymphoma w/TNF not significant	Wolfe and Michaud [97]
Cohort NDB USA	13,001	Range: 0.5–7.8	Malignancy incl lymphoma	Skin tumors ↑ No increase in solid tumors or lympho-proliferative malignancies	Wolfe and Michaud [98]
Cohort Sweden	53,076	Median: 4.8	Invasive melanoma, in situ melanoma, invasive cancers all sites	50 % ↑ in relative risk of invasive melanoma, no ↑ for in situ melanoma or invasive cancer all sites	Raaschou et al. [94]
Cohort Sweden	6366 (newly starting)	Up to 6 years	Cancers	No increase	Askling et al. [100]
Cohort Sweden	4160 pts	Range: 1–3+	Cancers	No increase	Askling et al. [99]
Cohort Denmark	9696	Mean: 2.9	Cancers	No increase for cancer overall but excess found for ovarian and colon	Dreyer et al. [103]
Cohort SABER USA	29,555 RA pts	Range: 0.5–1.5	Cancers	No increase	Haynes et al. [101]
Cohort USA and Canada	8458	Not specified	Hematologic malignancies and solid tumors	No increase	Setoguchi et al. [102]
RCT[a]	63 with 29,423 pts	Range: 0.5–3	Malignancy	No increase	Lopez-Olivo et al. [104]
RCT	9 with 3493 pts	Range: up to 1	Malignancy	Dose-dependent increased risk in malignancies	Bongartz et al. [95]
RCT	9 with 3316 pts	Range: up to 1	Malignancy	Trend to increased risk higher with etanercept (not significant)	Bongartz et al. [96]

[a]Randomized clinical trial

increased risk in malignancy (relative risk of 1.91) whereas etanercept usage resulted in a decreased risk (relative risk of 0.43) and infliximab was associated with a relative risk of 1.23. Beyond the first year follow up, the relative risks for all three were similar (0.80–0.83) [100]. Although no association was observed between solid tumors and lymphomas, an apparent increase risk in skin cancers, both nonmelanoma and melanoma skin cancer has been observed in two studies [94, 98].

While it is important to understand the potential risk of new malignancies, other questions such as whether anti-TNF treatment is associated with an increased risk of recurrent malignancies or associated with a worse prognosis of cancer that occurs either during or after treatment. For example, should physicians treat patients with anti-TNF who have a history of malignancy or if so how long should they wait following recovery from malignancy to start treatment? Data for two separate observational studies (German and British registries) did not demonstrate an increased risk of malignancy. Although in one study there was a slightly higher recurrence rate in those patients with prior malignancy in the anti-TNF treatment group as compared with patients treated with DMARDs, it was not statistically significant. In addition, patients without prior malignancy did not show an increased risk as compared with the unexposed patients [105]. While these data are encouraging, it should be noted that they are based on relatively small sample sizes most likely due to the reluctance of physicians to prescribe anti-TNF treatment to a patient with prior malignancy [86, 105, 106].

The conflicting results observed among these studies are most likely due to differences in study type (RCT vs. cohort), duration of exposure to anti-TNF treatment, composition of control or cohort group and sample sizes. Other limitations include the rarity of lymphomas and other cancers which complicate statistical analysis in that relatively small changes in the numerator can result in a major change in estimated risk [96]. Another complication is that lymphoma is increased in the general RA population [107]. In some of these studies it appeared that the increased risk observed after the first year was not maintained which could be due to differences in cancer detection not cancer causation [100]. Another bias is called "channeling" in which the patients with more severe disease are the ones that receive the anti-TNF therapy earlier and who may already be at a higher risk of developing lymphoma [98].

8.4 Pharmacovigilance, Risk Management and Regulatory Actions Taken

8.4.1 Detecting and Assessing Cancer Risks

As discussed previously, the intended pharmacology of some biopharmaceuticals can be expected to lead to an increased carcinogenic risk such as immunmodulators and growth factors. As animal models have a low predictive value for these biopharmaceuticals, it has been the recommendation by ICH S6(R1) that post marketing follow-up for biotechnologically-derived proteins may be more informative. In such situations the problem may be dealt with by including it in the Risk Management Plan (RMP) as a potential risk, sometimes with obligatory additional studies. Cancer risks have been associated with a considerable number of medicines. Amongst these are medicines used for the actual treatment of cancer itself, but also a large number of medicines for chronic diseases. TNF-inhibitors intended for diseases such as rheumatoid arthritis, erythropoietin used in cancer patients, insulin glargine or GLP-1-analogs in diabetes mellitus.

The concerns of a cancer risk can arise at any time before or after approval. In other cases concerns rise after adverse events have been spontaneously reported or published in the scientific literature, or when the results of interventional or observational clinical studies have been published.

Whatever the source, a suspicion of higher occurrence of malignant or premalignant disease needs further assessment. An evaluation of the strength of the evidence includes the plausibility, biological mechanisms, dose-response relation, strength of the association, time-to-onset, consistency, specificity [108]. When causality is considered possible and relevant, additional action may be needed to further study, minimise, and communicate the risk. The impact on the benefit-risk balance and public health in general must be assessed.

Assessing cancer risks in patients is often difficult. While time may be pressing, due to the severity of the adverse event, data are frequently not available or of insufficient quality. Often it takes several years before the first data becomes available. Pharmacoepidemiological studies, using observational data may suffer from possible confounding by indication or disease severity (e.g. in the study of insulin or erythropoietin and cancer risks). The choice of suitable reference groups can be challenging, or non-users of comparable disease severity may simply not exist in case of debilitating/life-threatening or orphan indications (e.g. in insulin, somatropin or TNF-inhibitors). Induction and latency times need to be distinguished, but are often impossible to separate [109].

8.4.2 Examples from Post-marketing Experience

We list here a few examples of biopharmaceuticals for which malignancies were reported post marketing in the recent past. The overview is based on the European Medicines Agency (EMA) or FDA safety labelling updates and European public assessment reports (EPAR) and illustrates the challenges associated with justifying regulatory action (e.g. updating the product information, including communication of the risk) based on often a limited number (but serious) case reports, or difficult to interpret and sometimes contradictory results from large and lengthy studies.

Calcitonine (long term use) Miacalcin (calcitonin-salmon) Injection and Nasal Spray).

On 19 July 2012, the EMA completed a review of the benefits and risks of calcitonin-containing medicines, concluding that there was evidence of a small increased risk of cancer with long-term use of these medicines. The Agency's Committee for Medicinal Products for Human Use (CHMP) recommended that they should only be authorised for short-term use in Paget's disease, acute bone loss due to sudden immobilisation and hypocalcaemia caused by cancer. The CHMP also concluded that the benefits of calcitonin-containing medicines did not outweigh their risks in the treatment of osteoporosis and that they should no longer be used for this condition [110, 111].

In March 2014 the FDA issued a label update on Malignancy: In a meta-analysis of 21 randomized, controlled clinical trials with calcitonin-salmon (nasal spray or investigational oral formulations), the overall incidence of malignancies reported was higher among calcitonin salmon-treated patients (4.1 %) compared with placebo-treated patients (2.9 %). Among the tumor types, basal cell carcinoma was the most common type of tumor. Other types included breast cancer and non-melanoma skin cancers. The malignancy risk in individual studies was generally not statistically significant; however in CT 320 [112], a large vertebral fracture prevention trial in postmenopausal women, a statistically significant increase in risk of malignancy was observed (Odds-ratio = 1.62, 95 % CI: 1.00, 2.61). There was no excess of malignancies with Miacalcin for treatment up to 6 months, while at longer treatment durations more malignancies were reported with Miacalcin treatment than with placebo. The FDA advised that the benefits for the individual patient should be carefully considered against possible risks.

Tumor necrosis factor (TNF) inhibitors include infliximab (Remicade), adalimumab (Humira/Trudexa) or certolizumab pegol (Cimzia), or with a circulating receptor fusion protein such as etanercept (Enbrel) [113–116].

In Jan 2005 the CHMP revised the product information for infliximab (Remicade; EU approval in 1999) and adalimumab (Humira, EU approval in 2003) to include details of the post-marketing experience on malignancies and lymphoproliferative disorders, including incidence. At that time possible risk for the development of lymphomas or other malignancies in patients treated with a TNF-antagonist could not be excluded.

In 2006, post marketing reports of hepatosplenic T-cell lymphoma (HSTCL) were identified for the anti-TNF agent Remicade (infliximab). Since launch in 1998 to about early 2006, six cases of hepatosplenic T-cell lymphoma had been reported in patients with Crohn's disease treated with infliximab. Five of them were in the age range of 12–19 years. All patients were on concomitant treatment with azathioprine or 6-mercaptopurine. Based on the data presented, a causal relationship of hepatosplenic T-cell lymphoma and infliximab therapy cannot be excluded. The relevant sections of the SmPC were updated to include the information on this finding, and a Direct Healthcare Professional Communication (DHPC) was sent out.

In May 2008 the product information of adalimumab (Humira/Trudexa) was revised related to the reports of rare cases of hepatosplenic T-cells lymphoma in patients treated with adalimumab. It was not considered that the reports of these rare cases of HSTCL alter the positive benefit/risk balance for adalimumab in the approved indications, when the PI is updated. Nevertheless, a DHPC was recommended. The target groups of prescribers for the DHPC should be the same as for the DHPC sent out about 2 years before for Remicade, due to the identification of HSTCL.

In June 2008 after three spontaneous reports of hepatosplenic T-cell lymphoma that prompted the update of section 4.4 and 4.8 of the SmPC of Humira®, a DHPC to alert prescribers about these very rare events was distributed. The CHMP Pharmacovigilance Working Party (PhVWP) suggested that a RMP of anti-TNF drugs should contain educational material aimed at the diagnosis of this extremely rare event.

Additionally, it was noted that this issue had been the subject of a recent communication from the FDA.

In November 2009 the FDA issued a label update (Boxed warning) on Malignancy: Lymphoma and other malignancies, some fatal, have been reported in children and adolescent patients treated with TNF blockers, of which Humira is a member.

In March 2010 based on a post marketing cumulative review of all malignancies in paediatric and young adult patients with infliximab a causal relationship between infliximab and the development of paediatric malignancies cannot be established. It is possible that concomitant exposure to other immunosuppressants and/or presence of underlying autoimmune diseases were contributory factors. Nevertheless, given its mechanism of action as TNF-blocking agent it cannot be excluded that infliximab may be also a contributing factor in the development of the observed malignancies.

In March 2010 cumulative reviews of the cases of leukaemia in adult and malignancies in paediatric patients reported with use of adalimumab did not allow establishing a causal relationship between the development of these malignancies and adalimumab. It is possible that concomitant exposure to other immunosuppressants and/or presence of underlying autoimmune diseases were contributory factors. Nevertheless, given its mechanism of action as TNF-blocking agent it cannot be excluded that adalimumab may be also a contributing factor in the development of the observed malignancies.

In November 2012 the cumulative review of registries, clinical trials and post marketing cases of Merkel cell carcinoma (MCC, or neuroendocrine carcinoma of the skin) coincident with infliximab or golimumab use identified 19 reports for infliximab and none for golimumab. All 19 reports were post marketing cases. No MCC cases were observed in registries and clinical trials. Of the 19 reports there were 2 fatalities reported in patients either taking multiple immunosuppressants concomitantly with infliximab or with limited information regarding medical history. Of the 19 reports, most of them had confounding factors (i.e. one or more risk factors for MCC such as prior immunosuppressant history, concomitant immunosuppressant therapies, and/or a history of malignancy) limiting the causality assessment with infliximab. Based on this review, MCC is considered causally associated with the use of infliximab, and a drug class effect to TNF inhibitors. Key factors supporting this conclusion include the biological plausibility based on immunosuppression by TNF-α inhibitors, the apparent sensitivity of MCC to immunosuppression, and the elevated reporting rate compared with the background rate of this type of cancer, all which suggest an association of MCC with this drug class. MCC is therefore added to section 4.8, with a frequency category of "Not known" for both infliximab and golimumab, as the frequency of the event cannot be estimated from the available data. The severity and seriousness of the event of MCC also justify its addition to section 4.4 warning prescribers that cases of MCC have been reported in patients treated with TNF blocker therapy and recommending periodic skin examination, particularly for patients with risk factors for skin cancer.

Based on the cumulative review of melanoma cases, coincident with infliximab or golimumab use it remains unclear whether a causal relationship exists between infliximab or golimumab use and the development of melanoma, however the possible contribution of infliximab or golimumab use to the risk cannot be excluded.

In Nov 2012 the cumulative search of the company clinical and post marketing databases for reports of possible Merkel Cell Carcinoma (MCC) or neuroendocrine carcinoma of the skin coincident with adalimumab therapy identified 15 reports of MCC. One report was from clinical trials and there were 14 postmarketing reports. Of the 14 post marketing reports, most of them had confounding factors and/or limited information to fully assess causality with adalimumab and 1 report had no confounding factors or alternative etiology reported. The 1 report of MCC from a clinical trial also had confounding factors. There were no fatalities due to MCC among the total of 15 reports. Although it is not clear whether the appearance of MCC in patients receiving adalimumab might be due to a number of factors such as other TNF inhibitor therapy, the underlying autoimmune diseases, sun exposure, the patient's age, or exposure to other non-biologic immunosuppressant therapy, the possible contribution of adalimumab use to the risk cannot be excluded.

In Jan 2013 for adalimumab (Humira) an index case of glioblastoma in a 28 year old female for the indication of psoriasis was reported. Unknown and very rapid onset of malignant grade 4 glioblastoma developed with fatal outcome 6 weeks after diagnosis. Up to Jan 2013 the MHRA had received the following UK cases; 5 cases of glioblastoma, 5 cases of brain neoplasm and 2 cases of brain neoplasm malignant. There is limited information on the onset from first dose in 5 of the other UK cases. 5 have an onset of >6 months. In two of the cases with onset >6 months it is possible that the neoplasm is a recurrence of previous brain cancer or has metastasised from a different primary malignancy. The MHRA have received the following Non-UK reports; 11 cases of glioblastoma, seven (7) cases of glioblastoma multiforme, fifteen (15) cases of brain neoplasm, and two cases of brain neoplasm malignant. The SmPC currently labels solid organ neoplasm including breast cancer, lung neoplasm and thyroid neoplasm.

"Other malignancies" is an important potential risk in the RMP of adalimumab. An increased risk of cancer is a known risk with all TNF inhibitors, although there is variation in which types are specifically mentioned in the product information. In the Pharmacovigilance Risk Assessment Committee (PRAC) it was discussed whether European registries are available to examine for further evidence, however none were known to be currently running except for the Rheumatoid Arthritis registry, which would be of limited use as it is open to all drug substances. It has therefore been decided to raise this issue.

In May 2013 the FDA issued a label update on Malignancies: The potential risk with the combination of azathioprine or 6-mercaptopurine and HUMIRA should be carefully considered.

A recently published Swedish cohort study Raaschou et al. [117] in 120 TNFi-treated and 120 biologics-naive individuals concluded that among patients with RA and a history of breast cancer, those who started TNFi-treatment did not experience more breast cancer recurrences than patients with RA treated otherwise.

Taken together the current pre and post-marketing results from studies, registries and spontaneous reporting is sometimes conflicting and many uncertainties remain regarding the contribution of risk factors and actual baseline risk with the patient under TNF inhibitor treatment (also see section E Table 1). Our post-marketing overview illustrates that past regulatory actions have been considered justified based the seriousness (although rare) of the cancer risk. Nevertheless, the remaining uncertainties present challenges for timing and content of risk minimisation and communication.

***Known risks*:**

Erythropoietin (Epogen/Procrit (epoetin alfa) and Aranesp (darbepoetin alfa)

In October 2006 section 5.1 of the SmPC of Aranesp was updated to modify the language relating to influence of darbepoetin alfa on tumour progression and survival for lymphoproliferative disease patients treated with darbepoetin alfa.

In July 2011 the FDA issued Safety Labeling Changes on risk on tumor progression or recurrence [118].

Somatropin (growth hormone) (high dose) risk on tumour progression or recurrence. In general, somatropin is contraindicated in the presence of active malignancy. Any pre-existing malignancy should be inactive and its treatment complete prior to instituting therapy with somatropin. Somatropin should be discontinued if there is evidence of recurrent activity. Since growth hormone deficiency may be an early sign of the presence of a pituitary tumour (or, rarely, other brain tumours), the presence of such tumours should be ruled out prior to initiation of treatment. Somatropin should not be used in patients with any evidence of progression or recurrence of an underlying intracranial tumour [119].

On 10 December 2010, the European Commission initiated a procedure under Article 20/Article 107 referral for somatropin-containing medicinal products and requested the CHMP to assess all the available data and its impact on the risk benefit balance for somatropin-containing medicinal products. The scope of the review was to assess the long-term safety of growth hormone treatments in light of the emerging safety data from the French 'Santé Adulte GH Enfant' (SAGHE) study. In particular the assessment regarded the potential increased risk of mortality due to diseases of the circulatory system, bone tumours and subarachnoid or intracerebral haemorrhage in children and when high doses are used. It looked at data on 10,000 adults who started treatment between 1985 and 1996, using a mandatory national registry. An analysis in approximately 7000 of those patients who were treated for growth

Table 8.2 Several patient based studies that investigated the cancer risk of insulin glargine

Type of study	Name 1st author	Year published	Country of patient database	Increased risk of malignancies
Cohort	Andersson [120]	2012	Denmark	No
Cohort	Colhoun [121]	2009	Scotland	No
Cohort	Currie [122]	2009	UK	No
Cohort	Hemkens [123]	2009	German	Yes
Cohort	Jonasson [124]	2009	Sweden	Only breast cancer
Cohort	Kostev [125]	2012	German	No
Cohort	Mannucci [126]	2010	Italian	Yes
Cohort	Ruiter [127]	2011	Netherlands	Only breast cancer
Rand. contr. trials	Home [128]	2009	–	No
Rand. contr. trials	Rosenstock [129]	2009	–	No

hormone deficiency and for gestational or idiopathic short stature showed a possible increased risk of mortality with somatropin compared with the general population. In particular, an increased risk of mortality due to bone tumours and cardiovascular events (such as bleeding in the brain) was seen. The risk appeared to be highest when doses higher than the ones approved were used.

Based on the evaluation of the currently available data and the scientific discussion within the Committee, the CHMP concluded that the benefit-risk balance of somatropin-containing medicines remains positive when used in the approved indications at the approved doses. However, to ensure that somatropin-containing medicines are used appropriately, the CHMP recommended that specific wording be included in the product information of all somatropin-containing medicines. In particular, the harmonised wording will emphasise that somatropin must not be used if there is any evidence of tumour activity, and that the maximum recommended daily dose should not be exceeded.

***Unknown risks*:**

Insuline glargine (also see section *C. Animal case studies a. Insulins and IGF-1*)

Conclusions of large population based studies were not consistent with each other. Problems with these patient based studies arise because investigators are fully dependent on the information that is provided in the database. Often patient information lacks for example, BMI, smoking habits, familial cancer incidence etc. while these factors might have a significant impact on the development of cancer (Table 8.2).

In addition, prescription information including the dose or duration of the treatment is not always known or taken into account. The follow-up duration of these studies is often short, less than 5 years, whereas it is doubtful whether this timeframe is long enough for a tumor to develop de novo. Further, there is the problem of causality; doctors might prescribe specific insulin analogues to patients with specific health related problems, so we might observe an increased carcinogenic risk of a certain treatment whereas the treatment itself does not cause such an effect. Lastly, some of these studies are severely criticized because of a lack of statistical analysis [130–132].

In May 2010, based upon the data that have become available since the granting of the initial Marketing Authorisation, the CHMP considered that the benefit-risk balance of Lantus remains positive, but determined that its safety profile should be closely monitored for the following reasons: Following the publication of four epidemiological studies on the risk of (breast) cancer with the use of insulin glargine in the journal *Diabetologia*, concerns were raised about the safety of insulin glargine in this respect. At the time of the first renewal (May 2010), three post-marketing pharmacoepidemiology studies were initiated by the MAH to further investigate the possible increased risk of cancer associated with the use of insulin glargine.

In June 2012 the results from the ORIGIN trial (Outcomes Reduction with an Initial Glargine INtervention) were published. This was a multinational 7-year randomized clinical study that investigated the effect of Lantus on cardiovascular (CV) morbidity and mortality in patients with pre diabetes (impaired fasting glucose [IFG], impaired glucose tolerance [IGT]) or early Type 2 diabetes mellitus (T2DM) who had evidence of CV disease.

Based on review of the data the CHMP concluded in December 2013 that a cancer relationship to insulin glargine was not demonstrated in any cancer subtype (e.g. breast, colon, prostate, lung), or for new or recurrent cancers, or deaths from cancer, over 6.2 years of median follow-up. Kaplan-Meier curves for the first cancer diagnosed during the trial, the first new cancer diagnosed, and death due to cancer were practically super imposable between the insulin glargine and standard care groups. Although this is important information, it has not been included in the SmPC because, (i) the design of the ORIGIN study was not anticipated in order to assess the risks of cancer (ii) it is questionable whether these results could be extrapolated to long-standing diabetes with high doses of insulin (iii) available epidemiological results regarding the risk of breast cancer for longer exposures to glargine are not fully consistent across different studies.

GLP1-agonists (exenatide (Byetta/Bydureon), approved in 2005/2012; liraglutide (Victoza), approved 2010]; lixisenatide (Lyxumia), approved in EU 2013; albiglutide (Eperzen), approved in 2014 by GSK). (*Also see section C. Animal case studies b. GLP1-agonists*) [61–64, 66].

In September 2011 (first renewal) based upon the data that have become available since the granting of the initial Marketing Authorisation, the CHMP considers that the benefit-risk balance of Byetta (exenatide) remains positive, but stated that its safety profile should be closely monitored for the following reasons: A number of safety issues have been identified for Byetta, in particular the potential association between exenatide and pancreatic cancer and thyroid neoplasms. The latter will be further investigated in a new epidemiological study. Also the possible drug interaction between exenatide and tacrolimus and exenatide and lamotrigine needs further evaluation.

References

1. International Conference on Harmonisation. ICH Guideline S1A: Need for Carcinogenicity Studies of Pharmaceuticals. November 1995. http://www.ich.org/fileadmin/Public_Web_Site/ICH_Products/Guidelines/Safety/S1A/Step4/S1A_Guideline.pdf. Accessed 23 Dec 2014
2. Van Oosterhout JP, Van der Laan JW, De Waal EJ, Olejniczak K, Hilgenfeld M, Schmidt V, Bass R (1997) The utility of two rodent species in carcinogenic risk assessment of pharmaceuticals in Europe. Regul Toxicol Pharmacol 25(1):6–17
3. Bronchud MH (2007) Molecular oncology. In: Meyers RA (ed) Cancer. From mechanisms to therapeutic approaches. Wiley-VCH Verlag, Weinheim, pp 3–54
4. Maher ER, Neumann HPH, Richard S (2011) von Hippel-Lindau disease: a clinical and scientific review. Eur J Hum Genet 19(6):617–623. doi:10.1038/ejhg.2010.175, Epub 2011 Mar 9
5. Somasundaram A, Ardanowski N, Opalak CF, Fillmore HL, Chidambaram A, Broaddus WC (2014) Wilms tumor 1 gene, CD97, and the emerging biogenetic profile of glioblastoma. Neurosurg Focus 37(6):E14. doi:10.3171/2014.9.FOCUS14506
6. Kinzler KW, Vogelstein B (1997) Gatekeepers and caretakers. Nature 386:761–763. doi:10.1038/386761a0
7. DeVries A, Van Oostrom CTM, Hofhuis FMA, Dortant PM, Berg RJW, DeGruijl FR, Wester PW, VanKreijl CF, Capel PJA, VanSteeg H, Verbeek SJ (1995) Increased susceptibility to ultraviolet-B and carcinogens of mice lacking the DNA excision repair gene XPA. Nature 77:169–173. doi:10.1038/377169a0
8. Melis JPM, Wijnhoven SW, Beems RB, Roodbergen M, vanden Berg J, Moon H, Friedberg E, van der Horst GTJ, Hoeijmakers JHJ, Vijg J, van Steeg H (2008) Mouse models for Xeroderma pigmentosum group A and group C show divergent cancer phenotypes. Cancer Res 68(5):1347–1353. doi:10.1158/0008-5472.CAN-07-6067
9. Robinson DE, MacDonald JS (2001) Background and framework for ILSI's collaborative evaluation program on alternative models for carcinogenicity assessment. International Life Sciences Institute. Toxicol Pathol 29(Suppl):13–19
10. Donehower LA, Harvey M, Slagle BL, McArthur MJ, Montgomery CAJ, Butel JS, Bradley A (1992) Mice deficient for p53 are developmental normal but susceptible to spontaneous tumors. Nature 355(6366):215–221
11. Hanahan D, Weinberg RA (2011) 2011 Hallmarks of cancer: the next generation. Cell 144(5):646–674. doi:10.1016/j.cell.2011.02.013
12. Tischler AS, Powers JF, Pignatello M, Tsokas P, Downing JC, McClain RM (1999) Vitamin D3-induced proliferative lesions in the rat adrenal medulla. Toxicol Sci 51(1):9–18
13. Silva Lima B, Van der Laan JW (2000) Mechanisms of nongenotoxic carcinogenesis and assessment of the human hazard. Regul Toxicol Pharmacol 32(2):135–143. doi:10.1006/rtph.2000.1427
14. Bronchud MH (2002) Is cancer really a 'local' cellular clonal disease? Med Hypotheses 59(5):560–565
15. Vogelstein B, Papadopoulos N, Velculescu VE, Zhou S, Diaz LA Jr, Kinzler KW (2013) Cancer genome landscapes. Science 339(6127):1546–1558. doi:10.1126/science.1235122
16. Pan Z-Z, Godwin AK (2007) Oncogenes. In: Meyers RA (ed) Cancer. From mechanisms to therapeutic approaches. Wiley-VCH Verlag, Weinheim, pp 55–114
17. EMA EPAR Kepivance 2005. http://www.ema.europa.eu/docs/en_GB/document_library/EPAR_-_Scientific_Discussion/human/000609/WC500040536.pdf
18. Sterner E, Meli L, Kwon SJ, Dordick JS, Linhardt RJ (2013) FGF-FGFR signaling mediated through glycosaminoglycans in microtiter plate and cell-based microarray platforms. Biochemistry 52(50):9009–9019. doi:10.1021/bi401284r
19. International Conference on Harmonisation. ICH Guideline S6(R1): Preclinical Safety Evaluation of Biotechnology-Derived Pharmaceuticals. June 2011. http://www.ich.org/fileadmin/Public_Web_Site/ICH_Products/Guidelines/Safety/S6_R1/Step4/S6_R1_Guideline.pdf. Accessed 23 Dec 2014

20. Vahle JL, Finch GL, Heidel SM, Hovland DN Jr, Ivens I, Parker S, Ponce RA, Sachs C, Steigerwalt R, Short B, Todd MD (2010) Carcinogenicity assessments of biotechnology-derived pharmaceuticals: a review of approved molecules and best practice recommendations. Toxicol Pathol 38:522–553. doi:10.1177/0192623310368984
21. Bugelski PJ, Volk A, Walker MR, Krayer JH, Martin P, Descotes J (2010) Critical review of preclinical approaches to evaluate the potential of immunosuppressive drugs to influence human neoplasia. Int J Toxicol 29(5):435–466. doi:10.1177/1091581810374654
22. International Conference on Harmonisation of Technical Requirements for Pharmaceuticals for Human Use. http://www.ich.org
23. Terrell TG, Green JD (1994) Issues with biotechnology products in toxicologic pathology. Toxicol Pathol 22(2):187–193
24. Inoue T (1998) Biotechnologically-derived pharmaceuticals in Japan: present and future prospects. In: Griffiths SA, Lumley CE (eds) Safety evaluation of biotechnologically-derived pharmaceuticals: facilitating a scientific approach. Kluwer Academic Publishers, Dordrecht, pp 51–63
25. Rosenblum IY, Dayan AD (2002) Carcinogenicity testing of IL-10: principles and practicalities. Hum Exp Toxicol 21(7):347–358
26. Ning S, Shui C, Khan WB, Benson W, Lacey DL, Knox SJ (1998) Effects of keratinocyte growth factor on the proliferation and radiation survival of human squamous cell carcinoma cell lines in vitro and in vivo. Int J Radiat Oncol Biol Phys 40(1):177–187
27. FDA Kepivance, Prescribing information 2004, 2013. http://www.accessdata.fda.gov/drugsatfda_docs/label/2013/125103s146lbl.pdf
28. EMA EPAR Neorecormon (2004) http://www.ema.europa.eu/docs/en_GB/document_library/EPAR_-_Scientific_Discussion/human/000116/WC500024975.pdf
29. FDA Epoietin Alfa, Prescribing information 1989. http://www.accessdata.fda.gov/drugsatfda_docs/label/2013/103234s5323lbl.pdf
30. Frasca F, Pandini G, Scalia P, Sciacca L, Mineo R, Costantino A, Goldfine ID, Belfiore A, Vigneri R (1999) Insulin receptor isoform A, a newly recognized, high-affinity insulin-like growth factor II receptor in fetal and cancer cells. Mol Cell Biol 19(5):3278–3288
31. Taguchi A, White MF (2008) Insulin-like signaling, nutrient homeostasis, and life span. Annu Rev Physiol 70:191–212
32. Pollak M (2008) Insulin and insulin-like growth factor signalling in neoplasia. Nat Rev Cancer 8(12):915–928. doi:10.1038/nrc2536
33. Pollak M (2012) The insulin and insulin-like growth factor receptor family in neoplasia: an update. Nat Rev Cancer 12(3):159–169. doi:10.1038/nrc3215
34. Malaguarnera R, Sacco A, Voci C, Pandini G, Vigneri R, Belfiore A (2012) Proinsulin binds with high affinity the insulin receptor isoform A and predominantly activates the mitogenic pathway. Endocrinology 153(5):2152–2163. doi:10.1210/en.2011-1843, Epub 2012 Feb 21
35. Manning BD, Cantley LC (2007) AKT/PKB signaling: navigating downstream. Cell 129(7):1261–1274
36. Taniguchi CM, Emanuelli B, Kahn CR (2006) Critical nodes in signalling pathways: insights into insulin action. Nat Rev Mol Cell Biol 7:85–96
37. Cheng Z, Tseng Y, White MF (2010) Insulin signaling meets mitochondria in metabolism. Trends Endocrinol Metab 21(10):589–598. doi:10.1016/j.tem.2010.06.005, Epub 2010 Jul 16
38. Schwartz GP, Burke GT, Katsoyannis PG (1989) A highly potent insulin: des-(B26-B30)-[AspB10, TyrB25-NH2]insulin(human). Proc Natl Acad Sci U S A 86(2):458–461
39. Milazzo G, Sciacca L, Papa V, Goldfine ID, Vigneri R (1997) ASPB10 insulin induction of increased mitogenic responses and phenotypic changes in human breast epithelial cells: evidence for enhanced interactions with the insulin-like growth factor-I receptor. Mol Carcinog 18(1):19–25
40. Glendorf T, Knudsen L, Stidsen CE, Hansen BF, Hegelund AC, Sorensen AR, Nishimura E, Kjeldsen T (2012) Systematic evaluation of the metabolic to mitogenic potency ratio for B10-substituted insulin analogues. PLoS One 7:e29198
41. Hansen BF, Kurtzhals P, Jensen AB, Dejgaard A, Russell-Jones D (2011) Insulin X10 revisited: a super-mitogenic insulin analogue. Diabetologia 54(9):2226–2231. doi:10.1007/s00125-011-2203-8, Epub 2011 Jun 3

42. Gallagher EJ, Alikhani N, Tobin-Hess A, Blank J, Buffin NJ, Zelenko Z, Tennagels N, Werner U, LeRoith D (2013) Insulin receptor phosphorylation by endogenous insulin or the insulin analog AspB10 promotes mammary tumor growth independent of the IGF-I receptor. Diabetes 62(10):3553–3560. doi:10.2337/db13-0249, Epub 2013 Jul 8
43. Ter Braak B, Siezen CL, Kannegieter N, Koedoot E, van de Water B, van der Laan JW (2014) Classifying the adverse mitogenic mode of action of insulin analogues using a novel mechanism-based genetically engineered human breast cancer cell panel. Arch Toxicol 88(4):953–966. doi:10.1007/s00204-014-1201-2, Epub 2014 Jan 25
44. Sciacca L, Cassarino MF, Genua M, Vigneri P, Giovanna Pennisi M, Malandrino P, Squatrito S, Pezzino V, Vigneri R (2014) Biological effects of insulin and its analogs on cancer cells with different insulin family receptor expression. J Cell Physiol 229(11):1817–1821. doi:10.1002/jcp.24635
45. Oleksiewicz MB, Bonnesen C, Hegelund AC, Lundby A, Holm GM, Jensen MB, Krabbe JS (2011) Comparison of intracellular signalling by insulin and the hypermitogenic AspB10 analogue in MCF-7 breast adenocarcinoma cells. J Appl Toxicol 31(4):329–341. doi:10.1002/jat.1590, Epub 2010 Oct 8
46. Dideriksen LH, Jørgensen LN, Drejer K (1992) Carcinogenic effect on female rats after 12 months administration of the insulin analogue B10 Asp. Diabetes 41:143A
47. Tennagels N, Welte S, Hofmann M, Brenk P, Schmidt R, Werner U (2013) Differences in metabolic and mitogenic signalling of insulin glargine and insulin aspart B10 in rats [corrected]. Diabetologia 56(8):1826–1834. doi:10.1007/s00125-013-2923-z, Epub 2013 May 8
48. Ter Braak B, Siezen CLE, Speksnijder EN, Koedoot E, van Steeg H, Salvatori DCF, van de Water B, van der Laan JW (2015) Mammary gland tumor promotion by chronic administration of IGF1 and the insulin analogue AspB10 in the $p53^{R270H/+}$WAPCre mouse model. Breast Cancer Res 17:14. doi:10.1186/s13058-015-0518-y
49. EMA (2001) Points to consider on the non-clinical assessment of the carcinogenic potential of human insulin analogues. http://www.ema.europa.eu/docs/en_GB/document_library/Scientific_guideline/2009/09/WC500003252.pdf
50. Sciacca L, Cassarino MF, Genua M, Pandini G, Le Moli R, Squatrito S, Vigneri R (2010) Insulin analogues differently activate insulin receptor isoforms and post-receptor signalling. Diabetologia 53(8):1743–1753. doi:10.1007/s00125-010-1760-6, Epub 2010 Apr 28
51. Kurtzhals P, Schaffer L, Sorensen A, Kristensen C, Jonassen I, Schmid C, Trub T (2000) Correlations of receptor binding and metabolic and mitogenic potencies of insulin analogs designed for clinical use. Diabetes 49(6):999–1005
52. Yehezkel E, Weinstein D, Simon M, Sarfstein R, Laron Z, Werner H (2010) Long-acting insulin analogues elicit atypical signalling events mediated by the insulin receptor and insulin-like growth factor-I receptor. Diabetologia 53(12):2667–2675. doi:10.1007/s00125-010-1899-1, Epub 2010 Sep 12
53. Sommerfeld MR, Muller G, Tschank G, Seipke G, Habermann P, Kurrle R, Tennagels N (2010) In vitro metabolic and mitogenic signaling of insulin glargine and its metabolites. PLoS One 5(3):e9540. doi:10.1371/journal.pone.0009540
54. Teng JA, Hou RL, Li DL, Yang RP, Qin J (2011) Glargine promotes proliferation of breast adenocarcinoma cell line MCF-7 via AKT activation. Horm Metab Res 43:519–523. doi:10.1055/s-0031-1280780, Epub 2011 Jul 19
55. Kellerer M, Haring HU (2001) Insulin analogues: impact of cell model characteristics on results and conclusions regarding mitogenic properties. Exp Clin Endocrinol Diabetes 109(1):63–64
56. Staiger K, Hennige AM, Staiger H, Haring HU, Kellerer M (2007) Comparison of the mitogenic potency of regular human insulin and its analogue glargine in normal and transformed human breast epithelial cells. Horm Metab Res 39(1):65–67
57. Pierre-Eugene C, Pagesy P, Nguyen TT, Neuille M, Tschank G, Tennagels N, Hampe C, Issad T (2012) Effect of insulin analogues on insulin/IGF1 hybrid receptors: increased activation by glargine but not by its metabolites M1 and M2. PLoS One 7(7):e41992. doi:10.1371/journal.pone.0041992, Epub 2012 Jul 26

58. Stammberger I, Bube A, Durchfeld-Meyer B, Donaubauer H, Troschau G (2002) Evaluation of the carcinogenic potential of insulin glargine (LANTUS) in rats and mice. Int J Toxicol 21(3):171–179
59. Stammberger I, Essermeant L (2012) Insulin glargine: a reevaluation of rodent carcinogenicity findings. Int J Toxicol 31(2):137–142. doi:10.1177/1091581811431111, Epub 2012 Jan 3
60. Ter Braak SJ, Wink S, Koedoot E, Pont C, Siezen CLE, Van der Laan JW, van de Water B (2015) Alternative signaling network activation through different insulin receptor family members caused by promitogenic antidiabetic insulin analogues in human mammary epithelial cells. Breast Cancer Res 17:97. doi:10.1186/s13058-015-0600-5
61. EMA EPAR Byetta 2006. http://www.ema.europa.eu/docs/en_GB/document_library/EPAR_-_Scientific_Discussion/human/000698/WC500051842.pdf
62. EMA EPAR Bydureon 2011. http://www.ema.europa.eu/docs/en_GB/document_library/EPAR_-_Public_assessment_report/human/002020/WC500108239.pdf
63. EMA EPAR Victoza 2009. http://www.ema.europa.eu/docs/en_GB/document_library/EPAR_-_Public_assessment_report/human/001026/WC500050016.pdf
64. EMA EPAR Lyxumia 2013. http://www.ema.europa.eu/docs/en_GB/document_library/EPAR_-_Public_assessment_report/human/002445/WC500140449.pdf
65. EMA EPAR Trulicity 2014. http://www.ema.europa.eu/docs/en_GB/document_library/EPAR_-_Public_assessment_report/human/002825/WC500179473.pdf
66. EMA EPAR Eperzan 2014. http://www.ema.europa.eu/docs/en_GB/document_library/EPAR_-_Public_assessment_report/human/002735/WC500165119.pdf
67. Sheiner LB, Steimer JL (2000) Pharmacokinetic/pharmacodynamic modeling in drug development. Annu Rev Pharmacol Toxicol 40:67–95
68. Danhof M, Alvan G, Dahl SG, Kuhlmann J, Paintaud G (2005) Mechanism-based pharmacokinetic-pharmacodynamic modeling-a new classification of biomarkers. Pharm Res 22(9):1432–1437, Epub 2005 Aug 24
69. Danhof M, de Lange EC, Della Pasqua OE, Ploeger BA, Voskuyl RA (2008) Mechanism-based pharmacokinetic-pharmacodynamic (PK-PD) modeling in translational drug research. Trends Pharmacol Sci 29(4):186–191. doi:10.1016/j.tips.2008.01.007, Epub 2008 Mar 18
70. Derendorf H, Meibohm B (1999) Modeling of pharmacokinetic/pharmacodynamic (PK/PD) relationships: concepts and perspectives. Pharm Res 16(2):176–185
71. Knudsen BL, Madsen LW, Andersen S, Almholt K, de Boer AS, Drucker DJ, Gotfredsen C, Egerod FL, Hegelund AC, Jacobsen SD, Moses AC, Mølck AM, Nielsen HS, Nowak J, Solberg H, Thi TD, Zdravkovic M, Moerch U (2010) Glucagon-like peptide-1 receptor agonists activate rodent thyroid C-cells causing clacitonin release and C-cell proliferation. Endocrinology 151(4):1473–1486. doi:10.1210/en.2009-1272, Epub 2010 Mar 4
72. Ponce RA, Gelzleichter T, Haggerty HG, Heidel S, Holdren MS, Lebrec H, Mellon RD, Pallardy M (2014) Immunmodulation and lymphoma in humans. J Immunotoxicol 11(1):1–12. doi:10.3109/1547691X.2013.798388
73. Salavoura K, Kolialexi A, Tsangaris G, Mavrou A (2008) Development of cancer in patients with primary immunodeficiencies. Anticancer Res 28:1263–1269
74. Weaver JL (2012) Establishing the carcinogenic risk of immunomodulatory drugs. Toxicol Pathol 40:267–271. doi:10.1177/0192623311427711
75. Cogliano V, Baan R, Straif K, Grosse Y, Secretan B, El Ghissassi F (2005) Carcinogenicity of human papillomaviruses. Lancet Oncol 6(4):204
76. Gaschen L, Schuurman HJ (2001) Ultrasound detection of non-Hodgkin's lymphoma in three cynomolgus monkeys after renal transplantation and cyclosporine immunosuppression. J Med Primatol 30(2):88–93
77. Kauffman HM, Cherikh WS, McBride MA, Cheng Y, Hanto DW (2006) Post-transplant de novo malignancies in renal transplant recipients: the past and present. Trans Internatl 19:607–620
78. Dempster AM (2000) Nonclinical safety evaluation of biotechnologically derived pharmaceuticals. In: El-Gewely MR (ed) Biotechnology annual review. Elsevier Science, Amsterdam, pp 221–258

79. Wordell CJ (1991) Biotechnology update. Hosp Pharm 26:897–900
80. Goldstein G (1987) Overview of the development of orthoclone OKT3: monoclonal antibody for therapeutic use in transplantation. Transplant Proc 19(2 Suppl 1):1–6
81. Swinnen LJ, Costanzo-Nordin MR, Fisher SG, O'Sullivan EJ, Johnson MR, Heroux AL, Dizikes GJ, Pifarre R, Fisher RI (1990) Increased incidence of lymphoproliferative disorder after immunosuppression with the monoclonal antibody OKT3 in cardiac-transplant recipients. N Engl J Med 323(25):1723–1728
82. Macsween KF, Crawford DH (2003) Epstein-Barr virus-recent advances. Lancet Infect Dis 3:131–140
83. Young LS, Rickinson AB (2004) Epstein-Barr virus: 40 years on. Nat Rev Cancer 4:757–768. doi:10.1038/nrc1452
84. Clarke J, Hurst C, Martin P, Vahle J, Ponce R, Mounho B, Heidel S, Andrews L, Reynolds T, Cavagnaro J (2008) Duration of chronic toxicity studies for biotechnology-derived pharmaceuticals: is 6 months still appropriate? Reg Toxicol Pharmacol 50:2–22. doi:10.1016/j.yrtph.2007.08.001
85. Bugelski PJ, Herzyk DJ, Rehm S, Harmsen AG, Gore EV, Williams DM, Maleeff BE, Badger AM, Truneh A, O'Brien SR, Macial RA, Wier PJ, Morgan DG, Hart TK (2000) Preclinical development of keliximab, a primatized™ anti-CD4 monoclonal antibody, in human CD4 transgenic mice: characterization of the model and safety studies. Hum Exp Toxicol 19:230–243. doi:10.1191/096032700678815783
86. Strangfeld A, Zink A (2010) Are we playing it safe? Tumor necrosis factor alpha inhibition and the risk of solid malignancies. Rheumatologist 1–7. http://www.the-rheumatologist.org/details/article/867933/Are_We_Playing_It_Safe.html
87. Carswell EA, Old LJ, Kassel RL, Green S, Fiore N, Williamson B (1975) An endotoxin-induced serum factor that causes necrosis of tumors. Proc Natl Acad Sci U S A 72(9): 3666–3670
88. Balkwill F (2009) Tumor necrosis factor and cancer. Nat Rev Cancer 9:361–371. doi:10.1038/nrc2628
89. Raaschou P, Simard JF, Neovius M, Askling J (2011) Does cancer that occurs during or after anti-tumor necrosis factor therapy have a worse prognosis? A national assessment of overall and site-specific cancer survival in rheumatoid arthritis patients treated with biologic agents. Arthritis Rheum 63(7):1812–1822. doi:10.1002/art.30247
90. Ferrajoli A, Keating MJ, Manshouri T, Giles FJ, Dey A, Estrov Z, Koller CA, Kurzock R, Thomas DA, Faderi S, Lerner S, O'Brien S, Albitar M (2002) The clinical significance of tumor necrosis factor-α plasma level in patients having chronic lymphocytic leukemia. Blood 100(4):1215–1219
91. Brown ER, Charles KA, Hoare SA, Rye RL, Jodrell DI, Aird RE, Vora R, Prabhakar U, Nakada M, Corringham RE, DeWitte M, Sturgeon C, Propper D, Balkwill FR, Smyth JF (2008) A clinical study assessing the tolerability and biological effects of infliximab, a TNF-α inhibitor, in patients with advanced cancer. Ann Oncol 19:1340–1346. doi:10.1093/annonc/mdn054
92. Baert FJ, D'Haens GR, Peeters M, Hiele MI, Schaible TF, Shealy D, Geboes K, Rutgeerts PJ (1999) Tumor necrosis factor alpha antibody (infliximab) therapy profoundly down-regulates the inflammation in Crohn's ileocolitis. Gastroenterology 116(1):22–28
93. FDA Etanercept, Summary basis of approval (1998) http://www.fda.gov/downloads/Drugs/DevelopmentApprovalProcess/HowDrugsareDevelopedandApproved/ApprovalApplications/TherapeuticBiologicApplications/ucm088697.pdf
94. Raaschou P, Simard JF, Holmqvist M, Askling J (2013) Rheumatoid arthritis, anti-tumor necrosis factor therapy, and risk of malignant melanoma: nationwide population based prospective cohort study from Sweden. BMJ 346:1–12. doi:10.1136/bmj.f1939, Published 8 April 2013
95. Bongartz T, Sutton AJ, Sweeting MJ, Buchan I, Matteson EL, Montori V (2006) Anti-TNF antibody therapy in rheumatoid arthritis and the risk of serious infections and malignancies. Systematic review and meta-analysis of rare harmful effects in randomized controlled trials. JAMA 295(19):2275–2285. doi:10.1001/jama.295.19.2275

96. Bongartz T, Warren FC, Mines D, Matteson EL, Abrams KR, Sutton AJ (2009) Etanercept therapy in rheumatoid arthritis and the risk of malignancies: a systematic review and individual patient data meta-analysis of randomised controlled trials. Ann Rheum Dis 68(7):1177–1183. doi:10.1136/ard.2008.094904
97. Wolfe F, Michaud K (2004) Lymphoma in rheumatoid arthritis. The effect of methotrexate and anti-tumor necrosis factor therapy in 18,572 patients. Arthritis Rheum 50(6):1740–1751. doi:10.1002/art.20311
98. Wolfe F, Michaud K (2007) Biologic treatment of rheumatoid arthritis and the risk of malignancy. Analyses from a large US observational study. Arthritis Rheum 56(9):2886–2895. doi:10.1002/art.22864
99. Askling J, Fored CM, Brandt L, Baecklund E, Bertilsson L, Feltelius N, Cöster L, Geborek P, Jacobsson LT, Lindblad S, Lysholm J, Rantapää-Dahlqvist S, Saxne T, Klareskog L (2005) Risks of solid cancers in patients with rheumatoid arthritis and after treatment with tumor necrosis factor antagonists. Ann Rheum Dis 64:1421–1426. doi:10.1136/ard.2004.033993
100. Askling J, van Vollenhoven RF, Granath F, Raaschou P, Fored CM, Baecklund E, Dackhammar C, Feltelius N, Cöster L, Geborek P, Jacobsson LT, Lindblad S, Rantapää-Dahlqvist S, Saxne T, Klareskog L (2009) Cancer risk in patients with rheumatoid arthritis treated with anti-tumor necrosis factor α therapies: does the risk change with the time since start of treatment? Arthritis Rheum 60(11):3180–3189. doi:10.1002/art.24941
101. Haynes K, Beukelman T, Curtis JR, Newcomb C, Herrinton LJ, Graham DJ, Solomon DH, Griffin MR, Chen L, Liu L, Saag KG, Lewis JD (2013) Tumor necrosis factor alpha inhibitor therapy and cancer risk in chronic immune mediated diseases. Arthritis Rheum 65(1):48–58. doi:10.1002/art.37740
102. Setoguchi S, Solomon DH, Weinblatt ME, Katz JN, Avorn J, Glynn RJ, Cook EF, Carney G, Schneeweiss S (2006) Tumor necrosis factor α antagonist use and cancer in patients with rheumatoid arthritis. Arthritis Rheum 54(9):2757–2764
103. Dreyer L, Mellemkjaer L, Andersen AR, Bennett P, Poulsen UE, Ellingsen TJ, Duijnhoven RG, Straus SM, Raine JM, de Boer A, Hoes AW, De Bruin ML (2013) Number of patients studied prior to approval of new medicines: a database analysis. PLoS Med 10(3):e1001407. doi:10.1371/journal.pmed.1001407, Epub 2013 Mar 19
104. Lopez-Olivo MA, Tayar JH, Martinez-Lopez JA, Pollono EN, Cueto JP, Gonzales-Crespo MR, Fulton S, Suarez-Almazor ME (2012) Risk of malignancies in patients with rheumatoid arthritis treated with biologic therapy: a meta-analysis. JAMA 308(9):898–908. doi:10.1001/2012.jama.10857
105. Strangfeld A, Hierse F, Rau R, Burmester G-R, Krummel-Lorenz B, Demary W, Listing J, Zink A (2010) Risk of incident or recurrent malignancies among patients with rheumatoid arthritis exposed to biologic therapy in the German biologics register RABBIT. Arthritis Res Ther 12(1):R5. doi:10.1186/ar2904, Epub 2010 Jan 8
106. Dixon WG, Watson KD, Lunt M, Mercer LK, British Society for Rheumatology Biologics Register Control Centre Consortium, Hyrich KL, Symmons DPM (2010) Influence of anti-tumor necrosis factor therapy on cancer incidence in patients with rheumatoid arthritis who have had a prior malignancy: results from the British Society for Rheumatology Biologics Register. Arthritis Care Res 62(6):755–763. doi:10.1002/acr.20129
107. Baecklund E, Iliadou A, Askling J, Ekbom A, Backlin C, Grannath F, Catrina AI, Rosenquist R, Feltelius N, Sundström C, Klareskog L (2006) Association of chronic inflammation, not its treatment, with increased lymphoma risk in rheumatoid arthritis. Arthritis Rheum 54(3):692–701. doi:10.1002/art.21675
108. Hill AB (1965) The environment and disease: association or causation? Proc R Soc Med 58:295–300
109. Stricker BH, Stijnen T (2010) Analysis of individual drug use as a time-varying determinant of exposure in prospective population-based cohort studies. Eur J Epidemiol 25(4):245–251. doi:10.1007/s10654-010-9451-7, Epub 2010 Apr 1
110. WHO. http://www.who.int/medicines/areas/quality_safety/safety_efficacy/pharmvigi/en/

111. Miacalcin team Novartis Pharmaceuticals Corporation *Available for Public Disclosure without Redaction* Miacalcin® (calcitonin-salmon) FDA Joint Reproductive Health Drugs and Drug Safety and Risk Management Advisory Committee Meeting on the Benefit/Risk of Salmon Calcitonin for the Treatment of Postmenopausal Osteoporosis – Briefing Book, 29 Jan 2013
112. Chesnut CH III, Silverman SL, Andriano K, Genant H, Gimona A, Harris S, Kiel D, LeBoff M, Maricic M, Miller P, Moniz C, Peacock M, Richardson P, Watts N, Baylink D (2000) A randomized trial of nasal spray salmon calcitonin in postmenopausal women with established osteoporosis: the PROOF Study. Am J Med 109(4):267–276
113. EMA EPAR Cimzia 2014. http://www.ema.europa.eu/docs/en_GB/document_library/EPAR_-_Procedural_steps_taken_and_scientific_information_after_authorisation/human/001037/WC500069736.pdf
114. EMA EPAR Humira. http://www.ema.europa.eu/docs/en_GB/document_library/EPAR_-_Procedural_steps_taken_and_scientific_information_after_authorisation/human/000481/WC500050869.pdf
115. EMA EPAR Remicade 2014. http://www.ema.europa.eu/docs/en_GB/document_library/EPAR_-_Procedural_steps_taken_and_scientific_information_after_authorisation/human/000240/WC500050890.pdf
116. EMA EPAR Enbrel 2014. http://www.ema.europa.eu/docs/en_GB/document_library/EPAR_-_Procedural_steps_taken_and_scientific_information_after_authorisation/human/000262/WC500027366.pdf
117. Raaschou P, Frisell T, Askling J (2014) TNF inhibitor therapy and risk of breast cancer recurrence in patients with rheumatoid arthritis: a nationwide cohort study. Ann Rheum Dis 0:1–7. doi:10.1136/annrheumdis-2014-205745
118. EMA EPAR Aranesp 2006. http://www.ema.europa.eu/docs/en_GB/document_library/EPAR_-_Procedural_steps_taken_and_scientific_information_after_authorisation/human/000332/WC500026145.pdf
119. EMA EPAR Somatropin referral 2011. http://www.ema.europa.eu/ema/index.jsp?curl=pages/medicines/human/referrals/Somatropin/human_referral_000287.jsp&mid=WC0b01ac05805c516f
120. Andersson C, Vaag A, Selmer C, Schmiegelow M, Sørensen R, Lindhardsen J, Gislason GH, Køber L, Torp-Pedersen C (2012) Risk of cancer in patients using glucose-lowering agents: a nationwide cohort study of 3.6 million people. BMJ Open 2(3). pii: e000433. doi:10.1136/bmjopen-2011-000433. Print 2012
121. Colhoun HM (2009) Use of insulin glargine and cancer incidence in Scotland: a study from the Scottish Diabetes Research Network Epidemiology Group. Diabetologia 52(9):1755–1765. doi:10.1007/s00125-009-1453-1, Epub 2009 Jul 15
122. Currie CJ, Poole CD, Gale EA (2009) The influence of glucose-lowering therapies on cancer risk in type 2 diabetes. Diabetologia 52(9):1766–1777. doi:10.1007/s00125-009-1440-6, Epub 2009 Jul 2
123. Hemkens LG, Grouven U, Bender R, Gunster C, Gutschmidt S, Selke GW, Sawicki PT (2009) Risk of malignancies in patients with diabetes treated with human insulin or insulin analogues: a cohort study. Diabetologia 52(9):1732–1744. doi:10.1007/s00125-009-1418-4, Epub 2009 Jun 30
124. Jonasson JM, Ljung R, Talback M, Haglund B, Gudbjornsdottir S, Steineck G (2009) Insulin glargine use and short-term incidence of malignancies-a population-based follow-up study in Sweden. Diabetologia 52(9):1745–1754. doi:10.1007/s00125-009-1444-2, Epub 2009 Jul 9
125. Kostev K (2012) Risk of breast cancer in patients on long-acting insulin analogues in comparison with those on human insulin. Diabetologia 55(5):1554–1555. doi:10.1007/s00125-012-2497-1, Epub 2012 Feb 19
126. Mannucci E, Monami M, Balzi D, Cresci B, Pala L, Melani C, Lamanna C, Bracali I, Bigiarini M, Barchielli A, Marchionni N, Rotella CM (2010) Doses of insulin and its analogues and cancer occurrence in insulin-treated type 2 diabetic patients. Diabetes Care 33(9):1997–2003. doi:10.2337/dc10-0476, Epub 2010 Jun 14

127. Ruiter R, Visser LE, van Herk-Sukel MP, Coebergh JW, Haak HR, Geelhoed-Duijvestijn PH, Straus SM, Herings RM, Stricker BH (2012) Risk of cancer in patients on insulin glargine and other insulin analogues in comparison with those on human insulin: results from a large population-based follow-up study. Diabetologia 55(1):51–62. doi:10.1007/s00125-011-2312-4, Epub 2011 Sep 29
128. Home PD, Lagarenne P (2009) Combined randomised controlled trial experience of malignancies in studies using insulin glargine. Diabetologia 52(12):2499–2506. doi:10.1007/s00125-009-1530-5, Epub 2009 Sep15
129. Rosenstock J, Fonseca V, McGill JB, Riddle M, Halle JP, Hramiak I, Johnston P, Davis M (2009) Similar risk of malignancy with insulin glargine and neutral protamine Hagedorn (NPH) insulin in patients with type 2 diabetes: findings from a 5 year randomised, open-label study. Diabetologia 52(9):1971–1973. doi:10.1007/s00125-009-1452-2, Epub 2009 Jul 16
130. Nagel JM, Mansmann U, Wegscheider K, Rohmel J (2010) Insulin resistance and increased risk for malignant neoplasms: confounding of the data on insulin glargine. Diabetologia 53(1):206–208. doi:10.1007/s00125-009-1535-0, Epub 2009 Sep 24
131. Pocock SJ, Smeeth L (2009) Insulin glargine and malignancy: an unwarranted alarm. Lancet 374(9689):511–513. doi:10.1016/S0140-6736(09)61307-6, Epub 2009 Jul 17
132. Bronsveld H, ter Braak B, Karlstad Ø, Vestergaard P, Starup-Linde J, Bazelier MT, De Bruin ML, de Boer A, Siezen CLE, van de Water B, van der Laan JW, Schmidt MK (2015) Treatment with insulin (analogues) and breast cancer risk in diabetics; a systematic review and meta-analysis of in vitro, animal and human evidence. Breast Cancer Res 17(1):100. doi:10.1186/s13058-015-0611-2

Chapter 9
Addressing Positive Findings in Carcinogenicity Studies

Todd Bourcier and Denis Roy

Abstract Cancer risk assessment is a major part of establishing the safety profile of investigational pharmaceuticals intended for chronic use. Because the nonclinical carcinogenicity assessment program is typically conducted in the later stage of clinical development and requires extensive resources, careful consideration must be given to its design, timing, conduct, interpretation, and reporting in order to protect patients. Positive results in carcinogenicity studies are not uncommon and can have a significant impact on the overall development program. Toxicologists interpreting carcinogenicity results and their relevance to human safety should use a multifaceted approach including statistical and exposure considerations. While for some investigational products the human relevance of a positive result can readily be discounted, others require follow-up assessments in order to characterize their relevance to human risk. Ultimately, even if a positive signal is concluded to be relevant to humans and occur at low or no margins of clinical exposures, its impact on the overall development program is shaped by the clinical indication, targeted population and its demonstrated therapeutic benefits. This chapter discusses some practical factors to consider when establishing whether a carcinogenicity study is 'positive', and establishing the path forward by determining the impact of positive findings on patient safety and the overall clinical development program, both from a Sponsor and FDA perspective.

Keywords Carcinogenicity • Positive • Tumor • Bioassay • Nonclinical

---

Disclaimer Sections of this article reflect the views of the author (T.B.) and should not be construed to represent FDA's views or policies

T. Bourcier, PhD
US Food and Drug Administration, Center for Drug Evaluation and Research, Silver Spring, MD, USA
e-mail: Todd.Bourcier@fda.hhs.gov

D. Roy (✉)
Nonclinical Drug Safety, SciLucent LLC, Vista, CA, USA
e-mail: d.roy@scilucent.com

M.J. Graziano, D. Jacobson-Kram (eds.), *Genotoxicity and Carcinogenicity Testing of Pharmaceuticals*, DOI 10.1007/978-3-319-22084-0_9

9.1 Introduction

Cancer risk assessment is a major part of establishing the safety profile of investigational pharmaceuticals intended for chronic use. Because the nonclinical carcinogenicity assessment program is typically conducted in the later stage of clinical development and requires extensive resources, careful consideration must be given to its design, timing, conduct, interpretation, and reporting in order to protect patients. As such, the International Conference on Harmonization (ICH) issued the S1B guidance in 1997 *Testing for Carcinogenicity of Pharmaceuticals* [1], which describes industry standards and study options to consider when an assessment of carcinogenicity is appropriate. In practice, the majority of the carcinogenicity assessments consist of a 2-year study in rats together with a 2-year study in mice or according to more recent industry trends, a 6-month study in transgenic mice.

The list of known or probable human carcinogens (IARC 1/2A) that were first detected or later observed in rodent bioassays is appreciable [2–4] and attests to the utility and sensitivity of the 2-year rodent bioassay. Yet, rodents are not perfect surrogates for cancer biology in humans, and the many limitations of the 2-year bioassay have often been debated in the literature [5–7]. Among the more resource-intensive limitations is the frequency of positive results that arise from the rat or mouse 2-year studies, which are subsequently perceived as not being relevant to human risk. Positive tumor results are not uncommon for 2-year rat and mouse studies. A retrospective analysis of 221 marketed pharmaceuticals submitted to German and Netherland regulatory agencies between 1980 and 1995 indicated that nearly half were associated with positive tumor results in rats, mice, or both [8]. A more recent analysis of pharmaceutical development programs submitted to the US FDA over a 20-month period that included both rat and mouse carcinogenicity assessments also revealed slightly more than half with a positive tumor outcome in at least one species (Table 9.1).

The implications and impact of a positive tumor outcome for a given pharmaceutical development program can vary greatly. While negative results are typically regarded as relative reassurance, tumor-positive results are typically met with a certain level of anxiety by sponsors as they might delay or preclude approval. The nature and mechanistic understanding of the rodent tumor response, interpreted within the clinical context of intended use of the pharmaceutical, greatly defines the path that a sponsor and the regulatory agency would ideally follow to address human risk and appropriately ensure patient safety. When the weight of evidence favors

Table 9.1 Two-species rodent carcinogenicity program outcomes reviewed by executive CAC (from Sept 2012 to May 2014)

		2-year mouse		hRas Tg mouse	
Study	Outcome	Positive	Negative	Positive	Negative
2-year rat	Positive	6	7	0	4
	Negative	1	11	0	4
	Total	7	18	0	8

drug approval, communication of the nonclinical carcinogenicity results to physicians and patients can take multiple forms and is an integral part of the package labeling ranging from a black box warning, restriction of clinical use, to merely descriptive results with little risk interpretation. In the US product label, a boxed warning, warnings and precautions (Section 5), Adverse Events (Section 6), Nonclinical Toxicology (Section 13), and Patient Counseling Information (Section 17) can provide relevant information regarding the potential carcinogenic effects of the drug. Similarly in the EU, Special Warnings and Precautions for Use (Section 4.4), Undesirable Effects (Section 4.8), Preclinical Safety Data (Section 5.3) and Other Conditions and Requirements of the Marketing Authorization (Annex II C) are provided in the Summary of Product Characteristics (SmPC).

In general, the totality of nonclinical data, with particular emphasis on the genotoxicity and carcinogenicity programs, together with attributes of the patient population and intended clinical indication, form the key database for establishing the path forward in terms of risk assessment and overall drug development strategy. When positive results are observed in a program, the type and nature of the changes observed inform the toxicologist on the potential mechanisms involved (genetic or epigenetic) during determination of their human relevance.

This chapter discusses some practical factors to consider when establishing whether a carcinogenicity study is 'positive', and establishing the path forward by determining the impact of positive findings on patient safety and the overall clinical development program, both from a Sponsor and FDA perspective.

9.2 Assessing a Potential Signal

From a practical drug development standpoint, the initial event surrounding a potential positive signal is typically encountered during interactions with the conducting laboratory Management, Study Director, Sponsor's study monitor and/or study pathologist. Unless concerns were already suspected prior to initiation or identified during the conduct of the carcinogenicity studies (e.g., class effects, pharmacology, positive or equivocal genotoxicity, pre-neoplastic lesions in chronic animal studies, palpable mass incidence on the study), most toxicologists are generally not anticipating any issues at this late stage of development, at least from the Sponsors perspective. On the other hand, the drug regulatory agencies might have a broader perspective and understanding given the multitude of programs that may be ongoing at any point in time, often from similar pharmacological classes of agents, and therefore have an opportunity to observe trends or detect signals before any individual Sponsor realizes it. This highlights the importance for the close interactions and communication between Sponsors and the appropriate drug regulatory agencies. The identification of a potential positive signal on any given program quickly becomes a focal point that can lead to a high level of anxiety, especially if the investigational new drug in development is a critical component of a portfolio both from a pharmaceutical industry and patient impact perspectives. When the team toxicologist is presented with the initial information, a wave of activities is typically necessary in order to properly respond. Communication in a timely manner both within and between organizations, including regulatory agencies is essential. One of the initial considerations when faced with a positive carcinogenicity signal is to determine whether any study design and/or data issues may be responsible for the results. This is a critical initial step and those who have been in this situation before, can attest to the importance of properly planning and executing a carcinogenicity program. Unfortunately, to this date, there are still Sponsors that put their program at risk by choosing not to follow standard industry practices or failing to reach out to regulatory agencies in order to gain concurrence on their program, including carcinogenicity study protocols.

The first challenge is to determine when enough evidence or information has been generated in order to establish if indeed the signal is a true signal. For example, receiving a call from the Study Director that it appears that the high dose animals have more palpable tumors than the control animals is cause for further investigation rather than a trigger to initiate broad internal and external communications on a positive signal. Conversely, receiving a draft carcinogenicity report that shows a statistically significant increase in many tumor incidences is cause for immediate and expedited attention. Not infrequently, a treatment-emergent tumor signal arises in the course of an ongoing rodent carcinogenicity study and it is most evident when the tumor is lethal because most or all of the early decedents would harbor the lethal tumor, but may also become apparent upon necropsy of early decedents that died from other causes. Also, positive palpation of masses, such as might occur with mammary neoplasms or lipomas, can be evidence of a treatment-emergent tumor signal without necessarily causing a rise in early mortality. Those examples would usually be fairly obvious for most toxicologists and raise immediate attention of a potential positive signal, but in some other cases it is less clear and can lead to differing opinions based on the situation. For example cases where the information is incomplete, preliminary (e.g. partial non-peer-reviewed draft pathology data), communicated verbally and/or coming from another source (e.g. literature, other drug in the same class). It is therefore imperative for the team toxicologist, along with the development team, to establish the foundation of the signal; namely, what data are available and what additional data may be required in order to determine the right course of action.

9.2.1 Study Design Considerations

An appropriately designed and executed carcinogenicity study can greatly help avoid pitfalls that can complicate or irreparably hinder adequate interpretation of study results. For example, the use of a poorly characterized excipient (unknown carcinogenic profile) or lack of appropriate control groups could cloud the relationship between an apparent positive tumor response and the investigational product being tested. Similarly, poor design, including inadequate dose selection leading to low drug exposure or to early loss of dose groups due to tolerability issues could potentially invalidate a study entirely. Given the complexity, scale, and resource requirements of carcinogenicity studies, mutual agreement on the study protocol with the relevant regulatory agencies prior to initiation of such studies is highly advisable. For this purpose, the FDA has established an Executive Carcinogenicity Assessment Committee (ECAC) that reviews study protocols submitted by sponsors as a Special Protocol Assessment [9]. The objective of the process is for the sponsor and ECAC to reach mutual consent on a study protocol that is most optimal in terms of achieving sufficient survival of animals for the dosing duration while maintaining sufficient exposure to the test drug, and valid control group designations for comparative evaluation. Mutual agreement on the study protocol and any ensuing major study design changes will generally ensure that the study's validity will not later be questioned, even if animal survival or other aspects of the study prove to be less than optimal. This is particularly relevant when a study conducted without prior agreement with ECAC reports no increase in tumor incidence, as differences in opinion on the adequacy of dose selection for that study may compel a recommendation or requirement that the study be repeated with more optimal doses. Similarly, a study conducted without ECAC agreement in which a significant loss of animals or early-termination in a significant number of groups (e.g., two out of three dose groups) might also compromise the study validity and require repeating the study with a modified protocol. In these two scenarios, it may be concluded that the carcinogenicity assessment was not robust due to insufficient exposure to drug or to inadequate survival of animals, respectively, which might have been avoided by review and concurrence with the protocol by the FDA ECAC prior to conducting the study. Accordingly, Sponsors should maximize the effective use of the toxicology and pathology resources available both from internal and contract research organizations and drug regulatory agencies when designing and executing their carcinogenicity program.

9.2.2 Assessing Study Data

The first descriptive evaluation in screening for potentially positive signals from a valid, acceptably executed rodent carcinogenicity study is to subject the tumor incidence data to a statistical analysis. Current FDA recommendations include routine analysis of trend and pairwise comparisons whereas the Society of Toxicologic Pathology (STP) currently recommends a preferred single statistical method, typically a trend test [10, 11]. Differences in survival of the control versus treated groups or, more commonly, among the test article treated groups can impact the incidence of both treatment emergent and spontaneous tumors in rodents. Therefore, statistical analysis of trend requires appropriate adjustment for survival, and this is recommended whether or not survival differences were observed in the study [10]. Also, as a general guide, the incidences of differing tumor types of similar histomorphogenic origin should be combined for statistical analysis [12]. Combining the incidences of tumor types should be conducted carefully as combining tumors with differing histomorphogenic origin may result in a positive statistical trend, which has no biological significance.

Following tests of trend, pairwise statistical comparisons of individual dose groups can allow identification of a no-effect dose and can importantly inform the risk assessment. The FDA currently follows the decision rules for tests of positive trend and pairwise comparisons as described in the 2001 CDER guidance [10] (Table 9.2). Note that the threshold for statistical significance is appropriately higher for tumor types with a common incidence than for those with a rare incidence.

In general, a drug is concluded to be 'positive' for inducing a given tumor type when statistical significance is achieved by both tests: the survival-adjusted trend test and an appropriate pairwise test between the control group and a treated group. While requiring statistical significance by both tests is expected to reduce the perceived high false positive rate of the bioassay, it is also expected to decrease the number of true positives and increase the chance for false negatives, which has implications for risk assessment. Therefore, there are circumstances where tumor responses that are significant by trend but not by pairwise testing should nonetheless be further scrutinized for potential relevance to human risk. A primary factor to consider is whether the numerical increase in tumor incidence follows a dose–response with due consideration given to any effect on survival. Equally relevant is whether a plausible relationship exists between the test drug's pharmacology or drug class and the type of tumor response observed. Historical incidence data on the tumor type in question from the most recent 2 or 3 years from the conducting laboratory helps interpretation of numerical increases in tumor incidence; however, further scrutiny of the tumor response is generally appropriate if a dose–response exists or biological plausibility is evident, or both.

9.2.3 Additional Considerations

As a sponsor and reviewer, assessing whether a potential positive carcinogenicity signal is related to the pharmaceutical can also be complemented by careful examination of existing and recent data from the development program for subtle changes and/or target organs that are now deemed relevant, and whether additional information on the drug class is available that identifies a risk that was unknown previously.

Table 9.2 Statistical decision rules for positive trend and pairwise comparisons in tumor incidences from 2-year rodent carcinogenicity studies

Statistical approach	Rare tumor Incidence ≤1 %	Common tumor Incidence ≥1 %
Trend (over ≥3 groups)	0.025	0.005
Pairwise	0.05	0.01

As such, it is good practice to review all relevant existing data in the context of the potential carcinogenicity signal to aid assessment of causality and especially to rule out any potential technical or scientific mistakes.

In some circumstances, a Sponsor might decide to conduct additional assessments from previously conducted nonclinical toxicology studies (for example, new histopathology sectioning, special stains, immunohistochemistry, on chronic studies) in order to further characterize the finding from the carcinogenicity program. Such analyses might help determine the relatedness of the finding to the pharmaceutical, or provide insight on a potential tumorigenic mode of action, or help identify an exposure-response relationship of toxicological findings relevant to the tumor type observed. However, numerous considerations need to be addressed prior to re-examining completed toxicology studies. For example, which studies or data will be re-evaluated? In cases where new data are generated, who will conduct the assessment (some investigational assessments would not be GLP-compliant)? Which studies should be reopened? Does the protocol (or report) of a finalized study need to be amended if the report is already filed to a regulatory agency? What is the impact of the additional work on the labeling language of approved drugs/biologics and their marketing application and/or active INDs? Do regulatory agencies outside of the US need to be notified and when? All these questions will become critical as toxicologists contemplate conducting additional assessments in light of a potential carcinogenicity signal and it can at times become overwhelming. In the authors' experience and practice, keeping in mind what is critical to protect patient safety and preserve compliance through timely communications with regulatory agencies is generally a good anchor point to guide the process, and manage all the stakeholders on the program impacted. Any new safety assessments conducted should definitely be addressed and documented by study protocol amendment and in cases where the Study Director is no longer with the testing laboratory, a new Study Director should be appointed and be part of the same study amendment. This is also true for any new or changed elements of the new activities that are different from the approved protocol and report (e.g. change in study pathologist, new methods applied on existing material/tissues). Dialog with the agency is also crucial at this stage and can be leveraged to clarify issues surrounding uncertainties around additional post-report finalization study assessments.

Assessing the impact on the labeling is typically part of an overall safety assessment process once the activities are completed and the new data evaluated. In cases where the new data significantly impact the risk to patients, a modification of the product labeling (or Informed Consent, Investigator's Brochures, or protocols for open INDs) would be warranted. Whether or not a product labeling change is or is not required, it is critical to keep the regulatory agencies informed of any new data that support the safety assessment of an investigational or marketed drug product.

Sometimes, complete data collection and review confirms that the carcinogenicity study is indeed positive and this is a stage where the team toxicologist will be confronted with the task of assessing its impact on the overall program and patient safety.

9.3 So You Have a Positive Signal! Now What?

9.3.1 Managing Sponsor's Responsibilities and Obligations

One of the critical roles of the toxicologist is to ensure that any signal detected in the program is appropriately documented, interpreted, contextualized, and reported in a timely fashion in the context of human risk assessment. Apart from the scientific aspect of the event, the toxicologist will quickly find himself/herself having to manage multiple stakeholders including the following depending on the organization and project complexity: project teams, senior management and executives, compliance, global safety, clinical, regulatory, legal, business partners and investors, alliance partners, marketing, portfolio management, media, and last but not least regulatory agencies. The overall goal should always be to ensure timely and proper risk communication across the board, but any toxicologist that has been through this situation before will likely agree that it is a daunting task.

As a sponsor toxicologist, the first communication line established is typically with the Contract Research Organization (CRO) and the immediate Management team. It is important to make sure that the message is clear and accurate, and toxicologists should avoid communicating conclusions too early and take into account existing data gaps and pending assessments. The main reason behind this approach is that until a clear understanding of the results has been established, releasing data or conclusions too early (internally or externally) can lead to confusion, miscommunications, or unnecessary efforts based on preliminary assessments. However, it is also recognized that timely communication with the regulatory agencies is a very important obligation, and ultimately communicating risks to patients is critical so toxicologists should exercise robust scientific judgment when evaluating a positive carcinogenicity signal and use a balanced approach between data generation, assessment, and communication.

In cases where additional information is needed, mapping the path forward is essential in ensuring that adequate risk assessment is conducted in order to effectively communicate in a timely manner. However in addition to defining the activities to address the positive carcinogenicity results, establishing a communication plan is often neglected and can lead to serious issues with stakeholders and potentially lead to inaccurate and improper information communication. A good communication plan is typically set up in multiple layers where the internal team and direct Management should be involved and openly engaged. Caution should be exercised as sometimes the project team tries to justify or establish the relevance of the signal (often thinking the signal is not relevant for their drug; for example common background tumor types involved in the positive study signal) before a careful examination has been conducted, which can lead to many issues. Once a clear discussion and communication plan has been established within the project team including the toxicologist, timely communication to the global safety, regulatory, compliance, clinical, legal and other internal departments as applicable, is essential and critical. The key for the toxicologist, when faced with a positive signal in a carcinogenicity study, is that the message is accurate, factual, and consistent across the organizations. The use of summary documents authored by the toxicologist and other relevant scientists can help standardize communication of the current knowledge to both internal and external recipients. While the internal communication is generally well underway, it is typically not long before other stakeholders, especially external collaborators or partners, request additional and more detailed information in order to comply with their own internal and external requirements and standard operating procedures (e.g. IND safety reporting, patient inform consents, labeling revision, media communication and press releases).

Another critical aspect that is often neglected or forgotten is how the toxicologist handles the communication with alliance partners and collaborators. There is often a high and urgent need from the alliance partners to get the information in a timely manner for internal and external purposes, sometimes including reporting obligations in other countries. In cases where data are incomplete or when new data or assessments are required, disagreements may sometimes threaten the timely reporting of the positive carcinogenicity results; the toxicologist remains the key ingredient at ensuring the information is communicated in a timely manner to protect patient safety. It is also not unusual, that different organizations may have different opinions on the interpretation of the results, regulations, or their obligations in addition to having different processes on how positive carcinogenicity results are handled. Therefore, the toxicologist should be aware of such differences early in the process (ideally at the inception of the project or as a minimum at the planning stage of the carcinogenicity program) in order to ensure that the activities are performed in a compliant manner for all parties involved.

9.3.2 *Agency Interactions: When and How*

If an early treatment-emergent tumor signal is suspected, there is an obligation of the sponsor to first assess whether the signal presents a significant risk to clinical trial participants such that a change in clinical investigations or safety information communication may be warranted, as discussed above. This is an obligation firstly to the patients exposed to the pharmaceutical, and is also compelled by the US code of federal regulations [13] that calls for submission of Investigational New Drug (IND) Safety Reports to the FDA and all participating investigators under the IND within 15 days of the sponsor making the determination that the finding indeed suggests a significant risk in humans exposed to the drug. It is the sponsor's responsibility to decide whether the finding meets this criteria (i.e., significant human risk suggested by data) or is too preliminary to interpret without replication or further investigation [14]. Generally, a finding that suggests a significant human risk warranting a 15-day IND safety report is one that necessitates a safety-related change in the clinical protocol, informed consent, investigator's brochure, or other safety aspect of clinical investigations.

On the other hand, assessing the significance of interim carcinogenicity results can be challenging. The interim incidence profile, proximity to clinical exposure, and the relationship between the potential tumor signal and the test drug's pharmacology are primary factors to consider when assessing a potential change in risk to the clinical population under study. Additional assessments as appropriate to understand the emerging signal may also prove informative in determining if a change in clinical safety investigations is necessary. For diverse reasons, Sponsors facing a positive signal in a carcinogenicity study may be more inclined to wait until more information is available before engaging with the regulatory authorities (e.g. waiting for the tumor data on the second species). However, there are clear advantages to informing the relevant regulatory authorities early of a significant carcinogenic signal from an ongoing or even completed carcinogenicity study as it is an opportunity to potentially resolve the safety issue prior to submitting a marketing application. In addition, the agency may have additional knowledge, not accessible to the Sponsor, which may enable guiding the Sponsor in focusing on specific key elements of the program. From a sponsor toxicologist's standpoint, the initiation of timely interactions with the regulatory agency will more than likely be severely challenging due to the presence of multiple stakeholders and diverging opinions and perspectives, and it is therefore the project toxicologist's duty to manage and advise wisely on the appropriate and relevant course of action for the development program.

9.3.3 Assessing Human Relevance

Assessing the relevance of rodent carcinogenicity study outcomes to human risk, whether tumors are observed or not, has been and will remain an ongoing subject of debate. Central to addressing positive results, regardless of signal strength, is determining the presence or absence of a safety margin, assessing the relevance of other toxicity occurring at the tumor-inducing dose, identifying key events that yielded the observed tumorigenic response in rodents, and addressing its relationship to human biology. Key considerations that are often encountered by sponsors and the FDA when addressing positive tumor findings in rodent bioassays for pharmaceuticals are discussed below.

9.3.3.1 Tumor Multiplicity and Cross-Species Signals

A tumor signal at the same site in both rodent species, or the presence of multiple tumor types in a single species is generally viewed as indicative of higher risk than a single tumor type found in one species or in one sex. This view is consistent with the observation that mutagenic chemicals and known human carcinogens are more often trans-species carcinogens than non-mutagenic chemicals [15]. Demonstration of trans-species carcinogenesis contributes importantly to IARC's classification of a chemical as a "probable/possible human carcinogen" [16], and is a compelling reason for conducting carcinogenicity studies in two rodent species, although extensive debates remain in the industry regarding this topic. Though the majority of pharmaceuticals are not mutagenic or genotoxic, positive tumor results in both species is considered more problematic than a single-species or single-sex finding, although a single-sex or single-species tumor signal is not a sufficient basis to minimize the potential importance of the finding.

9.3.3.2 Exposure Multiples and Safety Margins

Assessing risk from a clearly identified positive tumor finding in rodents can be greatly informed by comparing drug exposure achieved in the rodent study to the maximum human exposure intended for clinical use. Ideally, the rodent data is sufficiently clear to identify the lowest dose that induced the tumor (lowest adverse effect level [LOAEL]) and the highest dose that did not induce the tumor (the no-adverse effect level [NOAEL]). Exposure at these two doses, as assessed by area under the plasma concentration-time curve (AUC) measured at steady-state, are used together with the maximum clinical dose to define the safety margin. Specifically, the safety margin can be calculated by dividing the NOAEL exposure (AUC) in rodents by the maximum clinical exposure (AUC), which captures the gap between the exposure at the highest non-tumorigenic dose in rodents and human exposure as a fold-difference. Alternatively, and only when necessary, exposure defined as the dose corrected for body mass has been used. While exposure to the parent drug for small molecules is most commonly applicable, one needs to consider the contribution of metabolites to the tumor finding and whether metabolite exposure should be included in the comparison. Having defined a safety margin, the challenge becomes interpreting whether the safety margin is so large that one can reasonably conclude that the tumor finding in rodents would be of no consequence to human risk. For example, on one hand, if a safety margin is absent (if the exposure at the tumor-inducing dose in rodents is similar to clinical exposure), then other factors discussed in this chapter need to be considered very closely in assessing human relevance. Indeed, absent other relevant information, the absence of a safety margin greatly increases concern that a human risk exists, and must be addressed. On the other hand, a safety margin or LOAEL exceeding 25-fold clinical exposure (AUC) generally carries little concern for human risk, accompanied by certain assumptions discussed below. This interpretation is consistent with the ICH S1C guidance [17] that describes a 25-fold rodent-to-human exposure ratio as an acceptable pharmacokinetic endpoint in defining the high dose for rodent carcinogenicity studies. This 25-fold ratio has certain desirable attributes [17, 18] as it is sufficiently high to detect IARC 1 and 2A carcinogenic compounds, and sufficiently high to detect a majority of pharmaceuticals testing positive in rats. It can be inferred that if the high dose in a carcinogenicity study need not exceed a 25-fold exposure multiple, then any drug-related tumors observed at higher exposure multiples would present minimal concern for human risk. Not infrequently, exposure at one or more doses in carcinogenicity studies exceeds 25-fold human exposure, most often because the final clinical dose was not established at the time the studies were initiated and/or because of permissive pharmacokinetic and toxicological properties in rodents. The level of concern for human risk generally decreases as the safety margin increases. However, this approach should be carefully assessed when anti-drug antibodies are present in animals as they may impact positively (increase) or negatively (decrease) exposures which may not accurately reflect the true exposure to the drug (i.e. at the site of action), hence its carcinogenicity profile. Although more commonly observed following repeated administration of biological products, it has nonetheless been observed with peptides, and products using delivery systems, like complex polymers. Hence, risk assessment is greatly aided by additional consideration of the tumorigenic mechanism and by the presence or absence of confounding toxicity at the tumorigenic doses.

Currently, mouse carcinogenicity studies are more frequently being conducted in transgenic mouse models, particularly the p53+/− and Tg rasH2 [19, 20]. Different from 2-year studies conducted in wild-type mice, exposure multiples for either selecting the high dose in carcinogenicity studies or for assessing safety of positive findings are currently not applicable to transgenic mouse studies, which were validated based on maximum tolerated doses of test chemicals [21]. However, the use of exposure multiples rather than the MTD as a basis for dose selection for transgenic mouse carcinogenicity studies is still a subject of discussions and the conventional approach may change as knowledge evolves.

Conclusions based solely on safety margins from 2-year carcinogenicity studies makes several assumptions, notably that the tumorigenic mode of action, whether known or not, requires a threshold drug exposure below which the cancer risk is non-existent. Also, one assumes that a sufficient safety margin exceeds the importance of any differences in drug metabolism and pharmacologic action between the rodent species and humans, which could alter susceptibility to the tumor-inducing properties of the drug. However, when such assumptions are reasonable, concluding that human risk is minimal based on a safety margin or LOAEL that exceeds 25-fold clinical exposure, in the absence of confounding factors, is a rational and conservative risk assessment that is generally acceptable to regulatory agencies.

9.3.3.3 Maximum Tolerated Dose (MTD)

Assessing risk of drug-induced rodent tumors occurring at doses within 25-fold of clinical exposure can often be informed by determining whether the maximum tolerated dose was met or exceeded over the tumorigenic dose range. The value of this analysis is a presumption that tumors occurring at doses that exceed the MTD are of no consequence to human risk, whether the tumors emerged coincident with or as a cause of severe pathology [18, 22, 23].

In one case, when tumor emergence is coincident with but unrelated to severe toxicity, it is assumed that similarly toxic doses would not be given nor would be tolerated in clinical trials, thereby excluding tumorigenic risks at that specific dose.

In the other case, the severe metabolic and biochemical aberrations associated with exceeding the MTD are sometimes considered causative of the tumorigenic result, but are not seen at the tolerated dose in rodents nor expected at the still lower doses used in clinical trials, again excluding that dose and its associated risks from development. It should be noted here that for some disease conditions, such as cancer, a high degree of drug toxicity is tolerated or in fact intended, so adjustment must be made to the risk assessment for the potential lack of a safety margin in these cases.

To determine its potential impact on interpretation of positive tumor results, one must first recognize when a dose exceeds the MTD in a carcinogenicity study. A dose-related increase in mortality unrelated to tumor emergence provides the clearest indication that the MTD was probably exceeded in the carcinogenicity study. This can extend to morbidity necessitating euthanasia, early termination of a dose group, or cessation of dosing to aid survival of a dose group. However, excessive morbidity and mortality related to drug-emergent tumors is evidence that the MTD was not reached, and that other factors must be considered for assessing risk. If a dose is identified as having exceeded tolerability, then the next lowest dose without excess morbidity/mortality could be interpreted as having reached the MTD. Also, doses observed to exert a minimum toxic effect would be considered evidence of achieving an MTD, but not of exceeding the MTD, and again other factors would be considered for assessing risk if tumors are observed at that dose. Studies that evaluate a dose range that includes the MTD are generally considered a sensitive assessment of the carcinogenic potential of the test chemical [23].

Between consideration of exposure margins and tolerability of the tested doses, many tumor positive studies are interpreted as presenting minimal or no cancer risk to human patients, without the need for additional mechanistic studies. In such cases, results of the carcinogenicity studies and the fact they were conducted are disclosed in the drug's product label, in accordance with 21CFR 201.57 [24].

9.3.3.4 The Human Relevance Framework (HRF)

When a tumor signal is found, and the criteria in terms of risk assessment as described above are not met, the toxicologist and regulators will then be faced with the challenge of establishing the relevance of the tumor finding to human risks. Two critical assumptions have governed cancer risk assessment for many years, namely: In the absence of information to the contrary, (1) it is generally assumed that tumors observed in laboratory animals are predictive of human cancer, and (2) the mode of action (MOA) defined in laboratory animals also applies to humans [25]. In June 2001, the Risk Science Institute of the International Life Sciences Institute (ILSI RSI) formed a workgroup to examine these issues, with a focus on using MOA information to determine the human relevance of animal tumors. The workgroup developed a framework for MOA-based human relevance analysis (human relevance framework, HRF). Adoption of the HRF concept is now commonly used by many regulatory agencies and international organizations and is generally being recognized globally as a sound approach to assess human carcinogenicity risks, when needed. A significant number of publications are available on the topic [25–30]. Therefore, when a positive carcinogenicity signal is identified, human risk assessment relies on identifying a plausible MOA in animals and establishing whether or not the MOA is relevant to humans under the conditions of exposures.

The HRF outlines a four-part process, beginning with collection and assessment of data on the identified potential mode of action (MOA) in laboratory animals. A proposed MOA is then usually evaluated considering critical elements including: (1) is the weight of evidence sufficient to establish the MOA in animals; (2) are key events in the animal MOA plausible in humans; (3) taking into account pharmacokinetic and pharmacodynamic factors, is the animal MOA plausible in humans; all of which lead to a statement of confidence and implications for human risks.

Dose response and temporal relationships, along with analyses of the strength, consistency and specificity of key events, tumor responses, biological plausibility and coherence are critical elements that are also evaluated. Other potential MOAs also need to be accounted for in determining relevance to humans and is considered in the assessment of the weight of evidence as to the MOA and the level of confidence. The assessment should also identify inconsistencies and data gaps to explain the weight of evidence and the level of confidence, in addition to identifying additional data or studies that may be required [25].

In general, in the absence of information to the contrary, there is a presumption that animal tumors are relevant for human hazard assessment, and the animal MOA is presumed to describe processes in humans as well as in animals. No matter how well defined and fully analyzed, MOA information derived solely from animal studies does not permit definitive conclusions about human relevance or lack of relevance. Specifically, although an absence of human data permits an assumption of human relevance, conclusions about lack of human relevance depend in part on consideration of the potential applicability of the animal MOA to humans. In view of a myriad of different modes, developing criteria for determining what data or information is required and whether enough information exists to establish a particular MOA is difficult. However, once a MOA has been well delineated for one drug, data needs for verifying this mode for subsequent drugs working through the same MOA will usually be significantly reduced.

9.3.3.5 Assessing Human Relevance: An Integrated Approach

Multiple examples of positive carcinogenicity programs have been encountered and are available in the literature and most lead to a careful examination of the potential human relevance and their significance for risk assessment and communication.

Important considerations in assessing human relevance include the presence of tumors in rodent tissues with no anatomical human equivalent like the Harderian gland, Zimbal's gland, and forestomach. A positive finding in any of those tissues may raise questions about the significance of the tumors for human risks and hence, there is a general tendency to dismiss increased tumor findings in those tissues based solely on the absence of these organs/tissues in humans. This rationale is usually not appropriate as the presence of tumors could still be due to a relevant MOA involved in the carcinogenicity response and should therefore be carefully examined before being dismissed.

Additionally, treatment-related increases in certain common rodent tumor types are also often dismissed as irrelevant due to the common nature of the tumor in animals and sometimes combined with the rare occurrence in humans. For example, significant information is known concerning the MOA for DNA reactive genotoxic chemicals such as vinyl chloride and thorotrast, which are known to induce hemangiosarcomas in humans [31]. In contrast, numerous commercial chemicals that produce hemangiosarcomas in rodents act predominantly by non-DNA reactive, nongenotoxic, and proliferative mechanisms. The case of the peroxisome proliferator activated receptor (PPAR) agonists can illustrate the complexity and risks of potentially dismissing a common tumor type solely based on background incidence in a given species. Carcinogenicity data on multiple PPAR agonists (gamma, and dual) reported an increased incidence of hemangiomas and hemangiosarcomas in CD-1 and B6C3F1 male and female mice and hamsters. Of note, PPAR-gamma binding appears to be necessary, but not sufficient, to produce hemangiosarcomas, hence not all PPAR agonists were associated with increased tumor incidence in mice. The positive drugs were also associated with the development of tumors at multiples sites (liver, spleen, skin, urinary bladder, mammary gland, and adipose tissue) and angiomatous hyperplasia/angiectasis was also noted with several compounds in rodents and non-rodents. Despite the presence of a number of compound-specific initiating events (e.g., hemolysis, hypoxia, adipogenesis) that trigger nongenotoxic induction of hemangiosarcomas, there appears to be a common convergence of events including dysregulated angiogenesis and/or erythropoiesis that result from hypoxia and macrophage activation [31]. These lead to the release of angiogenic growth factors and cytokines that stimulate sustained endothelial cell proliferation that can lead to hemangiomas formation. Despite the advancement on the knowledge in possible MOAs involved, the species specificity of hemangiosarcomas is not well understood, especially in the case of nongenotoxic compounds. There are several lines of evidence that suggest that the mouse is more susceptible to the induction of hemangiosarcomas than humans, but additional work is still needed to determine if the differences established between rodents and humans represent sufficient qualitative differences to impact the overall risk assessment for these chemicals.

It is therefore imperative to consider the mechanism involved rather than solely the tumor type or incidence since the MOA might be shared between species, but targets potentially expressed in different tissues. Assessment of the cross-species sensitivity to the MOA becomes a critical component in determining whether an effect is relevant or not, so the location or tumor predisposition doesn't immediately allow the toxicologist to dismiss the finding.

Several rodent tumors induced by pharmaceuticals have published MOA's currently suspected as being species-specific [32–35]. When there is generally an established MOA that has been accepted as the basis for the absence of relevance to humans or a low probable relevance to humans, products with a carcinogenicity signal demonstrated to act through the same MOA only generally need to provide documentation of the key event involved in the response and its comparability to the established MOA.

For example, Leydig cell adenomas are frequently observed in nonclinical chronic toxicity studies and raised questions regarding their relevance for human risk assessment. A Working Group has established general guidances on rodent Leydig cell adenomas and their human relevance, based on multiple aspects of Leydig cell biology and toxicology including (1) control of Leydig cell proliferation; (2) mechanisms of toxicant-induced Leydig cell hyperplasia and tumorigenesis; (3) pathology of Leydig cell adenomas; (4) epidemiology of Leydig cell adenomas; and (5) risk assessment for Leydig cell tumorigens [36]. Occurrence of Leydig cell hyperplasia alone in test species after lifetime exposure to a chemical does not constitute a cause for concern in a risk assessment for carcinogenic potential, but early occurrence may indicate a need for additional testing. Seven hormonal modes of induction were studied, of which two (GnRH agonism and dopamine agonism) were considered not relevant to humans. Androgen receptor antagonism, 5α-reductase inhibition, testosterone biosynthesis inhibition, aromatase inhibition, and estrogen agonism were considered to be relevant or potentially relevant to human risks, but quantitative differences may exist across species, with rodents being more sensitive. It is therefore recommended that a margin of exposure approach be used for compounds causing Leydig cell adenoma by a hormonal mode that is relevant to humans [36].

Importantly, it is imperative for the toxicologist to keep in mind that even though current knowledge supports, through an MOA approach, the likely lack of relevance to human carcinogenicity risks, it is possible that additional information gathered at a later time or emerging data from human use may prove that the rodent finding was relevant after all.

Another example is the hyperprolactinemia-induced pituitary and mammary tumors by anti-dopaminergic compounds in rats that is generally considered as a rodent-specific phenomenon due to major differences in hormonal and reproductive functions between rats and humans [37]. These drugs act as nonselective dopamine receptors antagonists and interfere with all four dopamine pathways with therapeutic effects mediated through dopamine D2 and D4 receptors binding in the mesolimbic area, while side effects (including hyperprolactinemia) are mediated by D2 receptor blockade in the striatal area (extrapyramidal effects) and in the hypothalamic infundibular system [38]. However, in a response by Harvey [39], the author pointed out that the mechanism of prolactin-induced mammary gland carcinogenesis is actually very relevant to humans as evidenced by biological and molecular experimental data. Indeed, the collective epidemiological, pharmacological, clinical, and biological data on prolactin have shown that it is a tumor promoter that acts through similar mechanisms of action in both animals and humans. This case illustrates the complexity of determining human relevance of positive animal carcinogenicity findings and supports the fact that additional data is often times useful in establishing the potential relevance of a signal through the use of other animal models (transgenic, knock-outs), human data (natural knock-outs or deficiencies, excesses), and/or information from the human experience with the drug class. This is exemplified in the case study presented in Sect. 9.4.

9.3.3.6 Risks Communication

Regardless of the safety assessment and whether or not the data were sufficient or adequate to support the lack of human relevance, the product-specific assessment of carcinogenic potential is used to communicate risk and provide input to the risk management plan along with labeling proposals, clinical monitoring, post-marketing surveillance, or a combination of these approaches.

9.4 Case Studies

The following case study is one of many examples that illustrate how toxicologists may need to apply the scientific approaches, general concepts and principles outlined in this chapter when faced with positive results in their carcinogenicity program. The reader is reminded that this section will focus on the process involved in evaluating a positive signal, rather than discussing the interpretation of the data generated in the context of assessing human relevance of the rodent thyroid carcinogenicity findings, as opinions may differ at this time. As such, the following sections are meant to emphasize and outline the steps taken by the different organizations involved.

9.4.1 GLP-1 Receptor Agonists and Thyroid C-Cell Tumors

Glucagon-like peptide-1 (GLP-1) receptor agonists are currently approved for use in type 2 diabetes in multiple countries. Because of the fairly recent emergence of the thyroid c-cell carcinogenicity signal in both rodent species, they represent a good case scenario to illustrate how toxicologists may go about assessing a positive signal. The case scenario involves the first three US-approved GLP-1 agonists, namely Exenatide for Injection (Byetta®), Liraglutide (Victoza®), and Exenatide Once Weekly Injection (Bydureon®). Summaries of findings, approved labeling, and/or discussions at advisory committee meetings can be found on the FDA and European Medicine Agency (EMA) websites.

9.4.1.1 Exenatide Injection

Exenatide for injection (Byetta®) was the first GLP-1 agonist approved in the US in 2005 for the treatment of diabetes. Exenatide has a relatively short half-life, and is therefore marketed as a twice-daily peptide therapy. A complete nonclinical safety development program, including carcinogenicity assessments, was conducted with exenatide to support the marketing application for Byetta®. Carcinogenicity studies were conducted in CD-1 mice and Sprague–Dawley rats using once daily subcutaneous injection at doses representing exposure multiples of up to 95 times (in the mouse) and 5, 22, and 130 times (in the rat), respectively, the human systemic exposure. Of note, the rat and mouse carcinogenicity studies were conducted without FDA ECAC concurrence on the dose selection rationale provided by the Sponsor. It appears that the carcinogenicity studies were initiated by the Sponsor despite ECAC disagreement, and the studies were later found acceptable based on review of the achieved exposure margins, toxicity profile and survival data. In the mouse carcinogenicity study, exenatide treatment for 2 consecutive years, did not induce any increase in tumor incidence in any tissues and was therefore considered negative for carcinogenesis in that species. In rats, incidence of benign C-cell adenomas was increased among females at all doses (only statistically significant at the highest dose) compared to controls (Table 9.3). The increase in this common, age-related tumor type was accompanied by substantial increases in survival to 2 years among exenatide-treated rats. No statistically significant increase in exenatide-related tumors was observed when adjusting for survival.

Although non-statistically significant, the numerically increased incidence was considered toxicologically relevant because the incidence at the high dose group exceeded both concurrent and historical control rates, and the biological plausibility of a trophic effect of GLP1 agonism at the GLP1 receptors present on thyroid C-cells. A number of considerations discussed in this chapter, including the minor strength of the signal at the low and mid-doses, the high exposure margin at the high dose, and the common nature of the tumor in rats (and absence in mice), was considered sufficient to characterize the carcinogenic potential of exenatide without the need for additional studies. The product was approved with appropriate labeling to communicate potential risks:

> A 104-week carcinogenicity study was conducted in male and female rats at doses of 18, 70, or 250 mcg/kg/day administered by bolus SC injection. Benign thyroid C-cell adenomas were observed in female rats at all exenatide doses. The incidences in female rats were 8 % and 5 % in the two control groups and 14 %, 11 %, and 23 % in the low-, medium-, and high-dose groups with systemic exposures of 5, 22, and 130 times, respectively, the human exposure resulting from the maximum recommended dose of 20 mcg/day, based on plasma area under the curve (AUC). In a 104-week carcinogenicity study in mice at doses of 18, 70, or 250 mcg/kg/day administered by bolus SC injection, no evidence of tumors was observed at doses up to 250 mcg/kg/day, a systemic exposure up to 95 times the human exposure resulting from the maximum recommended dose of 20 mcg/day, based on AUC. (*Byetta Prescribing Label*; *02/15*).

Table 9.3 Incidence (%) of thyroid C-cell tumors in the rat carcinogenicity study with exenatide injection [40]

Rats	Males					Females				
Dose group, (mcg/kg/day)	0	0	18	70	250	0	0	18	70	250
Exposure ratio	NA	NA	5	22	130	NA	NA	5	22	130
C-cell adenoma (%)	12	15	15	23	15	8	5	**14**	**11**	**23**[a]
C-cell carcinoma (%)	0	1.5	0	0	0	0	0	0	0	0

[a]Statistically significant

Based on the evaluation of the exenatide nonclinical safety data available in multiple species, it appears that thyroid c-cell changes appear only in the rat carcinogenicity studies and was not present in any chronic repeated dose toxicology studies with exenatide in any species.

9.4.1.2 Liraglutide

The development program of liraglutide (Victoza®), another GLP-1 agonist product was also in development and was reviewed at an FDA Advisory Committee in 2009 and received US marketing approval in 2010. Liraglutide has an extended half-life compared to native GLP-1, providing an extended duration of exposure in most species, and hence supporting a once-daily dosing regimen. Carcinogenicity studies were conducted in CD-1 mice and Sprague-Dawley rats using once daily subcutaneous injection at doses representing exposure multiples of 0.2, 2, 10, and 45 times (in the mouse) and 0.5, 2, and 8 times (in the rat), respectively, the human systemic exposure. The investigational new drug was tested in a full range of genotoxicity and carcinogenicity assays. Although the genotoxicity program was negative, 2-year carcinogenicity studies performed in CD-1 mice and Sprague-Dawley rats identified carcinogenicity signals. A statistically significant carcinogenic signal in thyroid C-cells was identified in both sexes of both species (Table 9.4). Benign C-cell adenoma was observed in both sexes of mice starting at ten-times human exposure, with the next lowest dose exhibiting C-cell focal hyperplasia but without tumors. In addition, dorsal skin sarcomas at the injection site were significantly increased at the highest dose of 3 mg/kg/day liraglutide in

Table 9.4 Incidence (%) of thyroid C-cell tumors in the rodent carcinogenicity study with liraglutide [41]

Mice	Males					Females				
Dose group, (mg/kg/day)	0	0.03	0.2	1.0	3.0	0	0.03	0.2	1.0	3.0
Exposure ratio	N/A	0.2	1.8	10.0	45.0	N/A	0.2	1.8	10.0	45.0
C-cell focal hyperplasia (%)	0	0	1.5	17[c]	46[c]	0	0	10[b]	15[c]	29[c]
C-cell adenoma (%)	0	0	0	13[c]	19[c]	0	0	0	6[a]	20[c]
C-cell carcinoma (%)	0	0	0	0	0	0	0	0	0	3

Rats	Males				Females			
Dose group, (mg/kg/day)	0	0.075	0.25	0.75	0	0.075	0.25	0.75
Exposure ratio	NA	0.5	2	8	NA	0.5	2	8
C-cell focal hyperplasia (%)	22	28	40	48[a]	28	28	54[b]	48
C-cell adenoma (%)	12	16	42[c]	46[c]	10	27[a]	33[b]	56[c]
C-cell carcinoma (%)	2	8	6	14[b]	0	0	4	6

[a] $p<0.05$
[b] $p<0.01$
[c] $p<0.001$

male mice. Benign and malignant C-cell tumors were observed in both sexes of rats with a dose dependence starting at the lowest dose tested.

The FDA ECAC found both studies adequate based on the finding of treatment-emergent neoplasms in both species and in both sexes. Pre-malignant or neoplastic changes occurred at a drug exposure either below (in rats) or near (in mice) the clinical dose, providing no evidence of an adequate safety margin. The maximum tolerated dose was also not exceeded in these studies, with sufficient survival across all dose groups to study termination. Risk assessment therefore focused on an HRF approach, discussed below, that sought to address the tumorigenic mode of action and its potential relevance to human biology.

9.4.1.3 Human Relevance Assessment

One striking observation when reviewing the data in its entirety across the different GLP-1 agonists is the apparent weak signal in the exenatide injection program compared to the positive signal in the liraglutide programs. Given the similarity in the targeting properties (GLP-1 receptors) between the two different programs, one obvious difference is the exposure duration and/or profile to and the potency of the active ingredient in each program. While the exenatide injection carcinogenicity program used once-daily injection in both mice and rats, liraglutide program used a formulation and dosing regimen that lead to a more sustained exposure over any 24-h period in contrast to the exenatide injection (once daily regimen with a half-life of about 2 h would only provide total daily exposures of up to approximately 10–12 h). It is therefore imperative in interpreting the results between the different programs to consider the exposure profile and associated carcinogenicity results, hence the apparent differences between the short acting GLP-1 agonist (Byetta®) and the long-acting agonists (Victoza®).

Accordingly, the proposed MOA was based on general effects on cells that would be expected to occur via the GLP-1 receptor, due to the very high specificity of the agonists to the GLP-1 receptor, and the supporting exposure-response profile differences between the long-acting versus short-acting products.

The rat and mouse carcinogenicity studies with liraglutide showed no general overlap between tumor development and tissue GLP-1 receptor expression and the overall toxicology database did not suggest carcinogenicity or growth-promoting effects of liraglutide. The overall carcinogenicity data regarding long-acting GLP-1 agonists demonstrated that: (1) liraglutide showed a clear positive signal for C-cell tumors in both mice and rats; (2) liraglutide was also associated with positive increase of skin fibrosarcomas at doses ≥1 mg/kg/day (10× human exposure), and (3) NOAELs for thyroid c-cell tumors were observed at low exposure margins in mice (liraglutide only) and were not established in rats.

Given the carcinogenicity response, the type of tumors observed, the low or absent exposure margins, and the presence of tumors in multiple species and more than one program, establishing the human relevance of the findings became essential.

Accordingly, more than 30 nonclinical mechanistic studies were conducted with liraglutide to define the underlying mechanism, identify a biomarker, and to assess the human relevance of the carcinogenicity findings in the mouse and rat studies. In addition, intensive monitoring of calcitonin, a biomarker for C-cell mass and activation was performed in the clinical development program in more than 5000 subjects. The rodent findings also led to an increased focus on thyroid events and additional investigations were included in several of the intermediate and long-term clinical trials compared to what is standard for clinical development programs. The in vitro studies included studies characterizing GLP-1 receptor expression and function in rodents and man, receptor screening studies and mitogenicity studies in C-cells. The in vivo studies provided information from C-cell studies of varying duration in rodents and in non-human primates following up to 87-weeks exposure at more than 60-fold human exposure. The consolidated data from these studies apparently substantiated the sequence of events proposed by the Sponsor in the process leading to C-cell proliferation in rodents after long-term GLP-1 receptor agonist dosing and a proposed MOA was put forward.

1. GLP-1 receptor agonists bind to and activate GLP-1 receptors on C-cells;
2. GLP-1 receptor activation on C-cells induces calcitonin release;
3. Continued calcitonin secretion is followed by increased calcitonin synthesis;
4. Persistent stimulation of calcitonin synthesis is followed by C-cell hyperplasia in rodents;
5. Long-term C-cell hyperplasia may lead to C-cell neoplasia in rodents.

These identified key events in the MOA in rodents were compared to nonhuman primate and human data at both the qualitative and quantitative level to establish the relevance to humans. Based on the numerous studies conducted it was proposed that the rodent C-cell tumors induced by dosing of liraglutide were caused by a non-genotoxic, specific receptor-mediated mechanism to which rodents are particularly sensitive whereas non-human primates and humans are less sensitive.

Based on the carcinogenicity results and the additional HRF studies of liraglutide, the product was approved with a black box warning regarding thyroid c-cell tumors.

Warning: Risk of Thyroid C-Cell Tumors
See full prescribing information for complete boxed warning

- Liraglutide causes thyroid C-cell tumors at clinically relevant exposures in rodents. It is unknown whether Victoza causes thyroid C-cell tumors, including medullary thyroid carcinoma (MTC), in humans, as human relevance could not be determined by clinical or nonclinical studies (5.1).
- Victoza is contraindicated in patients with a personal or family history of MTC or in patients with Multiple Endocrine Neoplasia syndrome type 2 (MEN 2) (5.1).

Victoza Prescribing Label (04/13)

9.4.1.4 Exenatide Extended Release

Interestingly, following approval of liraglutide, rat carcinogenicity data from another long-acting GLP-1 agonist, namely Exenatide Extended Release (Bydureon®) became available. Exenatide Extended Release is a polymer-based slow-release formulation of exenatide (same active ingredient as Byetta®), allowing constant exposure to the active ingredient and a dosing regimen of once-weekly injection. The nonclinical development program leveraged the existing nonclinical safety program for Byetta® (containing the same active ingredient) and therefore a single rat carcinogenicity study was deemed sufficient to support the marketing application. The rat was selected not only based on technical feasibility (chronic injection of polymer-based product) but also based on the fact that it appeared to be the most relevant and sensitive species based on the results of the exenatide carcinogenicity program. The doses selected for the Exenatide Extended Release rat carcinogenicity study (0.3, 1.0, and 3.0 mg/kg every 2 weeks; 0.15, 0.5, and 1.5 mg/kg/week) were 2-, 9-, and 26-times higher than the proposed clinical dose based on AUC. The FDA ECAC concurred with the study design and proposed doses. The rat carcinogenicity study results (Table 9.5) indicated a statistically significant increase in incidence of C-cell adenomas at all doses in females and at 1.0 and 3.0 mg/kg in males.

Although the low dose male group did not show a statistically significant increase in C-cell adenomas, the incidence reported was above historical control data and was therefore considered positive, especially in light of the liraglutide carcinogenicity data. In addition, a statistically-significant increase in incidence of C-cell carcinomas was observed in the high dose group females (6 %) only, while incidences of 3 %, 7 %, and 4 % (non-statistically significant versus controls) were noted in the low, mid, and high dose group males. No further nonclinical studies were conducted to support its marketing application, and Exenatide Extended Released approval was significantly supported by the liraglutide HRF data and assessments.

Based on the carcinogenicity profile of Exenatide Once Weekly, the product was approved with a black box warning regarding thyroid c-cell tumors similar to liraglutide.

Table 9.5 Incidence (%) of thyroid C-cell tumors in the rat carcinogenicity study with exenatide once weekly [42]

Rats	Males					Females				
Dose group (mg/kg/2 weeks)	0	Pbo	0.3	1	3	0	Pbo	0.3	1	3
Exposure ratios	**NA**	**NA**	**2**	**9**	**26**	**NA**	**NA**	**2**	**9**	**26**
C-cell adenoma (%)	14	14	**29**	**46**[c]	**47**[c]	7	13	**31**[b]	**27**[b]	**30**[b]
C-cell carcinoma (%)	0	1	**3**	**7**	**4**	0	1	1	1	**6**[a]

Pbo placebo
[a]$p=0.06$
[b]$p<0.005$
[c]$p<0.001$

References

1. International Conference on Harmonization (ICH) S1B guidance. Testing for Carcinogenicity of Pharmaceuticals (1997) http://www.ich.org/. Accessed Feb 2015
2. Huff J (1999) Long-term chemical carcinogenesis bioassays predict human cancer hazards. Issues, controversies, and uncertainties. Ann N Y Acad Sci 895:56–79
3. Wilbourn J, Haroun L, Heseltine E, Kaldor J, Partensky C, Vainio H (1986) Response of experimental animals to human carcinogens: an analysis based upon the IARC Monographs programme. Carcinogenesis 7(11):1853–1863
4. Tomatis L, Bartsch H (1990) The contribution of experimental studies to risk assessment of carcinogenic agents in humans. Exp Pathol 40(4):251–266
5. Gold LS, Manley NB, Ames BN (1992) Extrapolation of carcinogenicity between species: qualitative and quantitative factors. Risk Anal 12(4):579–588
6. Alden CL, Lynn A, Bourdeau A, Morton D, Sistare FD, Kadambi VJ, Silverman L (2011) A critical review of the effectiveness of rodent pharmaceutical carcinogenesis testing in predicting for human risk. Vet Pathol 48(3):772–784. doi:10.1177/0300985811400445
7. Maronpot RR, Flake G, Huff J (2004) Relevance of animal carcinogenesis findings to human cancer predictions and prevention. Toxicol Pathol 32:40–48
8. Van Oosterhout JP, Van der Laan JW, De Waal EJ, Olejniczak K, Hilgenfeld M, Schmidt V, Bass R (1997) The utility of two rodent species in carcinogenic risk assessment of pharmaceuticals in Europe. Regul Toxicol Pharmacol 25(1):6–17
9. FDA Guidance for Industry – Special Protocol Assessment (2002) http://www.fda.gov/ucm/groups/fdagov-public/@fdagov-drugs-gen/documents/document/ucm080571.pdf. Accessed Feb 2015
10. CDER Guidance for Industry. Statistical Aspects of the Design, Analysis, and Interpretation of Chronic Rodent Carcinogenicity Studies of Pharmaceuticals (2001) http://www.fda.gov/downloads/drugs/guidancecomplianceregulatoryinformation/guidances/ucm079272.pdf. Accessed Feb 2015
11. The Society of Toxicologic Pathology's Recommendations on statistical analysis of rodent carcinogenicity studies (2002) Toxicol Path 30(3):415–418
12. Brix AE, Hardisty JF, McConnell EE (2010) Combining neoplasms for evaluation of rodent carcinogenesis studies. In: Hsu CH, Stedeford T (eds) Cancer risk assessment. Wiley, Hoboken, pp 699–715
13. IND Safety Reporting. Code of Federal Regulations, 21CFR312.32
14. CDER Guidance to Industry, Safety reporting requirements for INDs and BA/BE studies (2012) http://www.fda.gov/downloads/Drugs/…/Guidances/UCM227351.pdf. Accessed Feb 2015
15. Tennant RW (1993) A perspective on nonmutagenic mechanisms in carcinogenesis. Environ Health Perspect 101(Suppl 3):231–236
16. World Health Organization, International Agency for Research on Cancer (2006) IARC monographs on the evaluation of carcinogenic risks to humans, Preamble. Lyon
17. International Conference on Harmonization (ICH) S1C(R2) guidance. Dose selection for carcinogenicity studies of pharmaceuticals (2008) http://www.ich.org/. Accessed Feb 2015
18. Contrera JF, Jacobs AC, Prasanna HR, Mehta M, Schmidt WJ, DeGeorge J (1995) A systemic exposure-based alternative to the maximum tolerated dose for carcinogenicity studies of human therapeutics. Int J Toxicol 14;1:1–10. doi:10.3109/10915819509008676
19. Boverhof DR, Chamberlain MP, Elcombe CR, Gonzalez FJ, Heflich RH, Hernández LG, Jacobs AC, Jacobson-Kram D, Luijten M, Maggi A, Manjanatha MG, Benthem J, Gollapudi BB (2011) Transgenic animal models in toxicology: historical perspectives and future outlook. Toxicol Sci 121(2):207–233
20. Long GG, Morton D, Peters T, Short B, Skydsgaard M (2010) Alternative mouse models for carcinogenicity assessment: industry use and issues with pathology interpretation. Toxicol Pathol 38(1):43–50

21. Cohen SM (2001) Alternative models for carcinogenicity testing: weight of evidence evaluations across models. Toxicol Pathol 29(Suppl):183–190
22. Roth A, Kadyszewski B, Paulissen J, Weaver RJ (2007) Excess mortality in two-year rodent carcinogenicity studies. Toxicol Pathol 35:1040–1043
23. Haseman JK, Lockhart A (1994) The relationship between use of the maximum tolerated dose and study sensitivity for detecting rodent carcinogenicity. Fundam Appl Toxicol 22(3):382–391
24. Labeling requirements for prescription drugs and/or insulin. Code of Federal Regulations, 21CFR201 subpart B
25. Cohen SM, Meek ME, Klaunig JE, Patton DE, Fenner-Crisp PA (2003) The human relevance of information on carcinogenic modes of action: overview. Crit Rev Toxicol 33(6):581–589
26. Boobis AR, Cohen SM, Dellarco V, McGregor D, Meek ME, Vickers C, Willcocks D, Farland W (2006) IPCS framework for analyzing the relevance of a cancer mode of action for humans. Crit Rev Toxicol 36:781–792
27. Boobis AR, Doe JE, Heinrich-Hirsch B, Meek ME, Munn S, Ruchirawat M, Schlatter J, Seed J, Vickers C (2008) IPCS framework for analyzing the relevance of a noncancer mode of action for humans. Crit Rev Toxicol 38:87–96
28. Seed J, Carney EW, Corley RA, Crofton KM, DeSesso JM, Foster PM, Kavlock R, Kimmel G, Klaunig J, Meek ME et al (2005) Using mode of action and life stage information to evaluate the human relevance of animal toxicity data. Crit Rev Toxicol 35:663–672
29. Sonich-Mullin C, Fielder R, Wiltse J, Baetcke K, Dempsey J, Fenner-Crisp P, Grant D, Hartley M, Knaap A, Kroese D et al (2001) IPCS-Mode of action of chemical carcinogenesis. Regul Toxicol Pharmacol 34:146–152
30. Meek ME, Bucher JR, Cohen SM, Dellarco V, Hill RN, Lehman-McKeeman LD, Longfellow DG, Pastoor T, Seed J, Patton DE (2003) A framework for human relevance analysis of information on carcinogenic modes of action. Crit Rev Toxicol 33:591–653
31. Cohen SM, Meek ME, Klaunig JE, Patton DE, Fenner-Crisp PA (2009) The human relevance of information on carcinogenic modes of action: overview. Crit Rev Toxicol 33(6):581–589
32. Alison RH, Capen CC, Prentice DE (1994) Alison. Neoplastic lesions of questionable significance to humans. Toxicol Pathol 22(2):179–186
33. Proctor DM, Gatto NM, Hong SJ, Allamneni KP (2007) Mode-of-action framework for evaluating the relevance of rodent forestomach tumors in cancer risk assessment. Toxicol Sci 98(2):313–326
34. Maronpot RR (2009) Biological basis of differential susceptibility to hepatocarcinogenesis among mouse strains. J Toxicol Pathol 22(1):11–33
35. Silva Lima B, Van der Laan JW (2000) Mechanisms of nongenotoxic carcinogenesis and assessment of the human hazard. Regul Toxicol Pharmacol 32(2):135–143
36. Clegg ED, Cook JC, Chapin RE, Foster PM, Daston GP (1997) Leydig cell hyperplasia and adenoma formation: mechanisms and relevance to humans. Reprod Toxicol 11(1):107–121
37. Sistare FD, Morton D, Alden C, Christensen J, Keller D, Jonghe SD et al (2011) An analysis of pharmaceutical experience with decades of rat carcinogenicity testing: support for a proposal to modify current regulatory guidelines. Toxicol Pathol 39(4):716–744. doi:10.1177/0192623311406935
38. LaTorre D, Falorni A (2007) Pharmacological causes of hyperprolactinemia. Ther Clin Risk Manag 3(5):929–951
39. Harvey PW (2011) Prolactin-induced mammary tumorigenesis is not a rodent-specific response. Toxicol Pathol 39(6):1020–1022. doi:10.1177/0192623311141952
40. Food and Drug Administration. Byetta®: Summary Basis of Approval. http://www.accessdata.fda.gov/drugsatfda_docs/nda/2005/021773_Byetta_pharmr.PDF
41. Food and Drug Administration. Victoza®: Summary Basis of Approval. http://www.accessdata.fda.gov/drugsatfda_docs/nda/2010/022341s000TOC.cfm
42. Food and Drug Administration. Bydureon®: Summary Basis of Approval. http://www.accessdata.fda.gov/drugsatfda_docs/nda/2012/022200Orig1s000PharmR.pdf

Chapter 10
Nonclinical Strategies for Investigating Potential Tumor Signals Detected in Clinical Trials

Lorrene A. Buckley, Beatriz Silva-Lima, and Mark A. Tirmenstein

Abstract Potential signals of human carcinogenicity may arise in the course of clinical development or post marketing experience for a drug having shown a lack of evidence of a carcinogenic risk in nonclinical studies. It is always possible that, given the small numbers of patients in clinical trials, such signals may be due to chance or, for example, ascertainment bias; however, any signal of potential treatment-related malignancy must be evaluated and possible avenues of clinical and nonclinical investigation assessed. Investigations to characterize these signals should be considered thoughtfully, on a case-by-case basis, and grounded in scientific rationale. Given the relatively short time course of clinical development, tumor events are unlikely to have arisen de novo during the trial. Thus, potential mechanisms of tumor promotion and progression may also need to be considered. In this chapter, some nonclinical models to study tumor promotion and progression are discussed, and case studies are presented to illustrate various courses of follow-up investigations. Development and validation of innovative models for assessing tumor promotion and progression that are more human-based warrant further scientific investigation.

Keywords Nonclinical models • Carcinogenicity • Clinical tumor • Tumor progression • Tumor promotion

---

L.A. Buckley (✉)
Lilly Research Laboratories, Toxicology and Drug Disposition,
Eli Lilly and Company, Inc., Indianapolis, IN, USA
e-mail: BuckleyLA@lilly.com

B. Silva-Lima
iMED.ULisboa, Faculdade de Farmacia, Universidade de Lisboa,
Avenida Professor Gama Pinto, Lisboa 1649-003, Portugal
e-mail: beatrizlima@netcabo.pt

M.A. Tirmenstein
Drug Safety Evaluation, Bristol-Myers Squibb Company,
New Brunswick, NJ, USA
e-mail: mark.tirmenstein@bms.com

M.J. Graziano, D. Jacobson-Kram (eds.), *Genotoxicity and Carcinogenicity Testing of Pharmaceuticals*, DOI 10.1007/978-3-319-22084-0_10

10.1 Introduction

Assessment of potential carcinogenic risk to humans is an important component of safety assessment in drug development. Such assessments are multi-factorial and include: considerations of "on- and off-" target pharmacology; the ability of a drug to interact directly or indirectly with genetic material (in vitro and in vivo assays of genotoxicity); cellular or tissue toxicity that might presage the evolution of a carcinogenic effect (e.g., chronic inflammatory states, hyperplastic or dysplastic alterations, immunosuppression); and tumorigenic effects in traditional lifetime and/or alternative short-term carcinogenicity bioassays. Though long-term safety is evaluated in people during clinical trials and in post-market settings, carcinogenicity studies, per se, are not conducted in human subjects. Rather, lifetime bioassays in rodents are often employed as the "gold standard" to assess the potential of a compound to be carcinogenic. The lifetime nature of these studies allows for realization of all the stages of tumor development: initiation, promotion, and progression [52]. However, there may be translational issues in rodent bioassays models: analysis of study outcomes show that ~50 % of molecules tested are tumorigenic in lifetime rodent bioassays [21], though relatively few of the compounds (e.g., 20 % for pharmaceutical drugs) that are carcinogenic in rodents have been confirmed as human carcinogens [1]. Still, results from the lifetime studies in rodents have been shown to identify virtually all known human carcinogens [22].

In general, scientists and physicians take comfort in the belief that drugs which have shown no genotoxic or carcinogenic effects in nonclinical studies imply a low risk of carcinogenic risk in humans. However, potential signals of carcinogenicity may arise in the course of clinical development or post marketing experience for a drug having shown no potential for carcinogenic risk in nonclinical studies. This chapter will describe how potential clinical signals of carcinogenicity are evaluated and what nonclinical and clinical courses of investigation (additional to traditional genotoxicity and rodent carcinogenicity assays) might be considered to further understand and characterize possible human risk. Finally, case studies illustrative of potential human carcinogenicity signals in the (relative) absence of nonclinical signals will be discussed.

10.2 Detection and Investigation of Potential Clinical Signals of Cancer

Clinical trials are carefully designed to test whether a treatment is effective for a given disease state. However, even well-designed studies (such as a randomized, double-blind, placebo-controlled study) will have limitations with regard to controlling for all possible types of bias and the ability to detect a "true" rare event. Statistical power – the ability to detect a treatment- emergent event of a particular size or larger – will be limited by the number of patients in the trial. Small numbers of patients can lead to false signals due to chance or bias. "Detection" or "ascertainment" bias occurs when an event is more likely to be observed for a particular set of study subjects (e.g., prasugrel; see Sect. 10.3 below). For example, women taking an oral contraceptive will have more frequent cervical smears than women who are not on the pill and so are more likely to have any existing cervical cancer diagnosed. Not all potential signals raised in initial trials with limited numbers of patients are confirmed in larger trials or in post-marketing studies (e.g., orlistat; see Sect. 10.3 below) or with more extensive experience (e.g., angiotensin II AT-1 receptor blockers; [27]), while some potential signals are not detected until significant human post-marketing experience has accumulated (e.g., pioglitazone; see Sect. 10.3 below).

Monitoring for and interpreting adverse events in Phase 3 clinical trials is guided by what is known or reasonably anticipated about the therapeutic target and class, the biologic mechanism of action, previous (Phase 1 or 2) clinical experience, and the totality of the nonclinical safety assessment package. In the absence of evidence that a treatment may be associated with cancer – e.g., (depending on the circumstances of the product's development) previous experience with a related pharmacologic class having a known or suspected potential risk of human carcinogenicity and nonclinical signals – standard adverse event reporting would be performed in clinical trials. There would be no prospectively defined parameters for monitoring tumor events, and detection of a possible signal in the clinic would be unexpected. A report of cancer is a rare event in clinical trials and one for which there is a high degree of sensitivity. Minor numerical imbalances between control and drug treatment groups may represent a potential cause for concern. When one arm of a clinical trial suggests a possible increase in "all types" or a particular subtype of cancer, investigation of the potential for an identifiable cancer risk should be further evaluated.

10.2.1 Clinical Investigation

As mentioned, clinical investigators should be prepared to seek relevant clinical information for cancer events. There are many factors that might relate to carcinogenic risk including the baseline characteristics of the patient population (e.g., age, co-morbidities, smoking status, pre-existing conditions). As previously noted, while study groups are randomized for many baseline characteristics, one cannot account for every possible parameter; potential areas of bias need to be thoroughly investigated to rule out false positive signals. Link et al. [27] described the case of angiotensin II AT-1 receptor blockers (ARBs) in which a positive signal detected in a meta-analysis of five clinical trials was not recapitulated in a subsequent, much larger meta-analysis (31 clinical trials) which found no evidence of any excess site-specific cancer. (The larger meta-analysis results were supported by negative results in 19 mostly lifetime rodent bioassays for cancer when viewed en masse by FDA.)

10.2.2 Potential Relationship Between Pharmacologic Mechanisms and Tumorigenicity

Any assessment of carcinogenic potential begins with careful consideration of mechanism of action of an agent to hypothesize theoretical risks. There may be causes for concern for some pharmacological classes whether or not there is a signal in the nonclinical studies. An in-depth knowledge of the target biology will inform possible mechanisms by which a molecule may exert a carcinogenic effect. Such considerations form the basis for hypothesis generation that might be further tested in nonclinical studies.

There are several potential mechanisms of carcinogenesis. Genotoxic (DNA-reactive) agents interact directly and covalently with DNA to produce procarcinogenic mutations which lead to neoplastic transformation. Such agents are generally eliminated from consideration early in drug development. Epigenetic mechanisms of carcinogenesis (not associated with direct effects on the DNA sequence) can be quite complex in nature and may involve sustained adaptive effects and/or disruption of endocrine, paracrine, nervous, and immune systems; it is these types of pathways that are typically of interest for drugs in development [40].

The accumulated experience with rodent carcinogenicity studies makes evident that several pharmacological classes are associated with the development of rodent tumors at specific organs or tissues. Underlying mechanisms of rodent tumorigenicity may include the persistent stimulation of the involved target or of target-associated cell cascades or suppression of defense mechanisms against tumor cells. The potential relevance of these mechanisms for humans needs to be thoroughly understood in terms of whether tumorigenesis derives from the pharmacologic mode of action (e.g., is "target-related") and under which conditions (e.g., threshold doses, treatment schedules, exposure duration). The human relevance of animal data is dependent on multiple factors, including species specificities in: disposition and metabolic pathways; receptor levels, isoforms, polymorphisms; cross-talk between signal transduction pathways. Other sources of inter-species variability include regulatory pathways of cell repair, proliferation and death, and differences in patterns of cell/tissue adaptation or potentiation of responses during chronic exposures to chemicals. Differences in stem cell populations which are targets for neoplastic transformation across species also may influence receptor-mediated carcinogenic responses [2]. Therefore, extensive knowledge on the primary and secondary targets for any active substance, the associated cellular mechanisms, and their similarities and differences between rodents and humans is fundamental for understanding possible human risk posed by any rodent tumorigen. When the pharmacological mechanism responsible for the tumorigenesis in rodents is considered of possible/plausible relevance to humans, a potential concern with the pharmacological class may emerge even in the absence of positive results in rodents for some member(s) of such class.

Examples of rodent tumorigenesis triggered by certain pharmacological classes of drugs, several of which have been shown to have potential relevance to humans, are described below.

10.2.2.1 Receptor-Mediated Tumorigenesis

The pharmacologic activity of many drugs is mediated through binding to cellular receptors, which may result in proliferative responses at target tissues, and which, in turn, has been hypothesized to be associated with tumorigenesis. Hormones or hormone analogues may constitute the active substance of a medicinal product directly interacting with the respective receptor, or hormones may be indirectly modulated by an active substance. Agents acting at the parathyriod hormone, calcitonin, and dopamine receptors may be associated with a hormonally-mediated mechanism of tumorigenesis.

Calcitonin (CT) is a hormone produced by the C cells of the thyroid and participates in the regulation of blood calcium levels and calcium mobilization from the bone. Mainly, CT acts by decreasing calcium in the blood and inhibiting bone resorption. Salmon calcitonin (sCT) is a calcitonin analogue which was indicated for the chronic treatment of osteoporosis and reduction of fracture-associated pain. Unlike human calcitonin which acts specifically at the CT receptor, sCT has affinity to other receptors of the CT receptor family like calcitonin gene related peptide (CGRP) and amylin receptors. Levels of CT receptors are elevated in different tumors/tumor cells, and CT has been proposed to mediate tumor progression [42, 53]. In fact, a small but significant increase in post-marketing reports of tumors (different types) was reported in patients treated with sCT, and the use of injectable sCT-containing products has been restricted to shorter-term indications [16].

Dopamine antagonists (used as neuroleptics, for example) induce hyperprolactinemia through inhibition of the inhibitory action of dopamine on prolactin secretion in the hypothalamus. In line with this activity, mammary tumors are commonly observed in rats with this pharmacological class. However, those tumors have been considered non-relevant for humans based on the difference in prolactin physiology in rodents and humans. An increased incidence in breast tumors has not been established in patients receiving dopamine antagonists in clinical trials. Some relatively recent publications have raised the concern of a possible increase of breast cancer in patients treated with neuroleptics, however the available information remains inconclusive [51].

The pharmacologic actions of insulin analogues are mediated through insulin receptor (IR) stimulation, but the insulin hormone may also have proliferative effects which are triggered by its activation of insulin-like growth factor receptors (IGFR). These proliferative effects induced by insulin(s) in tissues expressing the IGFR may suggest a theoretical cause for concern even when no tumors (e.g., mammary tumors) are observed in chronic rodent studies.

10.2.2.2 Immunosuppressive Agents

There is compelling evidence that the immune system can identify and destroy nascent tumors and thereby function as a primary defense against cancer [41]. Particular immune cell types, effector molecules, and pathways can sometimes collectively function as extrinsic tumor suppressor mechanisms. Therefore, chronic disturbance of the immune system may reduce immune surveillance and possibly allow tumor growth. Although immunosuppression is a recognized risk factor for human carcinogenesis, rat carcinogenicity testing results with immunosuppressive agents do not reliably reflect this human risk [5], thus potentially yielding "missed signals" for human carcinogenicity. The increase in cancer risk after transplantation has been widely documented for different immunosuppressive regimens with, for example, cyclosporin, azathyoprin, methotrexate, and tacrolimus. Immunomodulatory therapies against several chronic diseases with autoimmune etiology, like rheumatoid arthritis, multiple sclerosis, and psoriasis may potentially be associated with risk of tumorigenicity. Even in the absence of a positive rodent carcinogenicity study or evidence in clinical trials for an individual immunosuppressant, the tumorigenic risk posed by immunosuppression itself should be considered when establishing the benefit: risk ratio of a drug.

In conclusion, there are several pharmacological classes associated with a consistent pattern of tumorigenesis. For most of these classes, it may be plausible that the mechanism of tumorigenesis is associated to their primary or secondary target receptors and may also occur in humans, but the human relevance in the real life conditions of clinical use should be clarified. Agents that are growth factors, hormones or which cause hormonal stimulation, or suppress the immune system may be potentially associated with carcinogenic risk. In these cases, any potential safety risks need to be evaluated and identified. Additionally, for new molecular entities with innovative mechanisms of action, a deep and thorough knowledge of the target biology and associated pharmacologic cascades is needed regarding hypotheses of any carcinogenic risk.

10.2.3 Nonclinical Safety Assessment

Prior to Phase 3 clinical trials, the genotoxic potential of the drug has been characterized; however, typically, the nonclinical carcinogenicity studies have not been completed and in some cases, the results of chronic toxicology studies may not yet be available [10]. The results of subchronic (≤3 months) or chronic (≤9 months) repeated dose toxicology studies together with a thorough knowledge of the pharmacologic attributes of the product may help identify potential risk factors, e.g., hyperplasia, endocrine activity, immunotoxicity.

Given the relatively short course of time during clinical development – many trials are less than 6 months in duration – tumor events are unlikely to have arisen de novo during course of the trial. If tumors are observed in clinical trials, they are likely pre-existing, although the possibility of some drug-associated mechanism of tumor promotion or progression should also be considered. The rodent carcinogenicity studies may provide some evidence of tumor promotion or progression activity if a treatment-related increase in the incidence or onset of spontaneously developed tumors occurs. Other nonclinical models which have been used to investigate the promotion/progression are described below.

10.2.3.1 Investigative Nonclinical Models

Experimental animal models for examining tumor promotion are limited and have not been validated for predicting human risk. The most widely used model is the 2-stage tumor promotion model in which rodents are first dosed with a tumor initiator for a short period of time followed by the putative tumor promoter or a vehicle control. The promotion phase of the model usually involves several months of repeated dosing following initiation. After several months of promoter treatment, the incidence and severity of tumors are scored and compared between rodents receiving the hypothesized tumor promoter and those receiving the vehicle control. Compounds that significantly increase the incidence and severity of tumors present over those treated with the vehicle control are classified as tumor promoters.

The 2-stage tumor promotion model has been used to examine tumor promotion in several different tissues including skin, liver, stomach, urinary bladder, and pancreas [9]. In most cases, the initiating agent is a genotoxic agent, while the tumor promoter is usually non-genotoxic and acts by increasing cellular proliferation. Cell proliferation can occur in response to tissue injury or in response to a mitogenic stimulus.

Although widely used in the literature, several problems have been identified with this model. It has been demonstrated in many cases that promoters themselves can act as complete carcinogens and induce tumors without the need of prior exposure to an initiating agent. Therefore, it is difficult to definitively distinguish agents as tumor initiators or promoters [9]. There is also considerable variation in how the 2-stage tumor promotion models are conducted with regards to types of initiating agents used, timing of administration of initiators and putative promoters, lag time for the development of tumors, and differences in species, sex and strains used. These differences make standardization of the model for routine tumor promoter identification difficult. Finally, no effort has been made to assess the model with respect to its utility in identifying potential tumor promoters in humans. Sodium ascorbate for example is a urinary bladder tumor promoter in rats [9], but there is no indication that sodium ascorbate functions as a urinary bladder tumor promoter in humans [23]. Therefore the model has not been validated with regards to identifying human tumor promoters and is not currently required by health authorities for assessing the safety of drugs. Since an agent that increases cell proliferation will test positive as a potential tumor promoter, it has been argued that use of the 2-stage tumor promotion model is unnecessary, and that agents that induce cell proliferation by any mechanism (cell toxicity or mitogenic stimulus) can be identified by screening for the presence of hyperplasia in chronic toxicity studies [9].

Assays to characterize the potential association between a drug and tumor growth and progression in animals are commonly conducted in pharmaceutical biology laboratories to enable discovery of potential oncolytic therapies. Typically, in vitro systems with human tumor cell lines are employed to measure inhibition of cell proliferation. Subsequently, in vivo studies are conducted to evaluate tumor growth using mouse xenograft models. Although not standardly employed to evaluate drugs and enhancement of tumor cell proliferation and growth, studies of this type were conducted with prasugrel and dapagliflozin; methodological details are described in Buckley et al. [4] and Reilly et al. [37]. In brief, the methods used were as follows:

For the in vitro cell proliferation experiments, human tumor cell lines of interest (related to potential concerns identified in the clinical trials) are plated in complete media containing 10 % Fetal Bovine Serum (FBS). Following an overnight incubation, the 10 % FBS-containing media is removed and replaced with serum-free media, thus "starving" the cells of growth factors. The drug is added after 24–48 h of starvation at concentrations approximating human plasma levels or some multiple thereof. Cell proliferation is then measured (the WST-1 cell proliferation assay was used in the prasugrel studies) and statistically analyzed. Both a negative (vehicle) control and a positive control (10 % FBS) are included.

For the mouse xenograft studies, the drug is repeatedly administered to nude mice harboring human tumor xenografts derived from subcutaneous implants of human tumor cell lines. Drug administration commences when tumors reached approximately 100 mm^3. The dosing period is based on the anticipated growth rates of the tumors; the study should be terminated when tumors reach a predetermined size considered to cause undue stress to the animals. Rates of growth (size and estimated volume) of the xenografts are statistically compared with those of the vehicle control treated animals throughout the study. Weights and volumes of excised tumors at study termination are also measured and analyzed. Excised tumor tissue can be preserved for possible histopathological examination.

As mentioned, these models are not well-validated for use in the evaluation of tumor promotion and progression for human risk assessment of drugs, and such studies should be considered on a case-by-case basis considering the biology of the molecule and other relevant factors. While good animal models suitable to assess carcinogenic risk associated with impaired immune function are lacking, the potential risk to humans for this class of agents is recognized and accepted, regardless of whether any tumor imbalances are observed in the clinical program. If the known biology or pharmacology of an agent is not sufficient to explain tumor findings in clinical trials, alternative models should be considered. Given that significant uncertainty exists with such systems based on limited knowledge of clinical relevance and inexperience with alternative assessments, it is hoped that technological and scientific advances in the areas of systems biology, computational biology, predictive in silico approaches, molecular biology and genomics centered in human-based systems will provide better tools for identifying true human carcinogens (e.g., [3]).

10.3 Case Studies

Drugs characterized by a lack of evidence of carcinogenic potential in nonclinical assessments have been associated with varying degrees of evidence of potential clinical tumor signals and consequent courses of signal evaluation. The case studies below provide some examples and illustrate various pathways of investigation that may be considered in evaluating potential cancer risks.

10.3.1 Dapagliflozin

Dapagliflozin is a selective sodium-glucose co-transporter 2 (SGLT2) inhibitor that is currently marketed in the United States (Farxiga, AstraZeneca), EU (Forxiga, AstraZeneca), and several other countries for the treatment of Type 2 diabetes. SGLT2 inhibitors promote urinary glucose excretion thereby reducing hyperglycemia and lowering glycosylated hemoglobin (HbA_{1c}) levels. SGLT2 is selectively expressed in the proximal tubule of the kidney [7, 25, 39] and therefore SGLT2 inhibitors would be expected to have no direct pharmacologic effects on other tissues. Humans with functional mutations in SGLT2 exhibit familial renal glucosuria but are largely asymptomatic. SGLT2 inhibitors are insulin-independent and can therefore be used in conjunction with many existing antidiabetic therapies. Other benefits of the SGLT2 inhibitors include low risk of hypoglycemia, weight loss due to loss of calories in the urine, and decreases in blood pressure due to the diuretic effects of urinary glucose excretion.

Dapagliflozin was subjected to a standard battery of nonclinical toxicology testing as required by regulatory health authorities [43]. There was no evidence that dapagliflozin was genotoxic as assessed by in vitro bacterial reverse-mutation assays. There was also no evidence that dapagliflozin was clastogenic in vivo in rats after daily dosing of 200 mg/kg dapagliflozin for 1 month (Cmax exposures ≤544× maximum recommend human dose [MRHD]). Dapagliflozin did not induce an increase in unscheduled DNA synthesis in male rats or induce clastogenicity in bone marrow micronucleus studies when tested to the maximum tolerated dose (700 mg/kg) required by ICH guidelines. In the 2-year rodent carcinogenicity assays, there was no evidence that dapagliflozin induced tumors or shortened the latency period for tumor development. Dapagliflozin did not increase the incidence of spontaneous background tumors in either the mouse or rat carcinogenicity studies. Mammary gland tumors, for example, are a common background lesion in female rats. There was no evidence that dapagliflozin increased the incidence of mammary gland tumors over those observed in control rats. Therefore, based on the weight of the nonclinical evidence, there was no evidence suggesting that dapagliflozin was a carcinogen.

However, during the dapagliflozin global clinical program (2011 cutoff), a numerical imbalance was observed in the number of urinary bladder tumors detected in dapagliflozin treated patients (nine patients with tumors—incidence rate 0.15) vs. those receiving placebo (one patient with tumor—incidence rate 0.03) [24]. The overall incidence of malignancies and unspecified tumors was balanced between dapagliflozin and placebo groups. Both the sponsor and the FDA indicated that there was no nonclinical evidence for dapagliflozin inducing cancer. The FDA acknowledged that there was a possibility that glucuosuria (with increased urination) and related genito-urinary infections in dapagliflozin-treated patients may contribute to a detection bias for cases of urinary bladder cancer. According to this hypothesis, patients treated with dapagliflozin may have a greater detection of hematuria compared to those treated with placebo requiring further work-up and a higher rate of cancer diagnosis. Despite the lack of any nonclinical signal, the imbalance in urinary bladder tumors still raised regulatory concerns. There was no evidence that dapagliflozin acted as tumor initiator, but it was suggested that dapagliflozin may act as a urinary bladder tumor promoter or enhance tumor progression. A reassessment of previously conducted toxicity studies and additional nonclinical studies were conducted to evaluate whether dapagliflozin may act as a tumor promoter or enhance tumor progression.

Metabolism: The primary human metabolite of dapagliflozin is dapagliflozin 3-O-glucuronide (D3OG) [43]. This metabolite is a stable non-reactive ether glucuronide that lacks pharmacologic activity. The 3-O-glucuronide is formed at a lower rate in preclinical species, but plasma concentrations comparable to or higher than human exposures were achieved at the doses of dapagliflozin used in the toxicological assessment of dapagliflozin. Non-human species also form a pharmacologically active O-deethylated metabolite, which is only a minor metabolite in humans [36]. The possibility that urinary metabolites of dapagliflozin may be involved in the induction of bladder cancer was also assessed. No unique human dapagliflozin metabolites were found in the urine. Bridging studies were used to extrapolate urinary exposures to dapagliflozin metabolites. Similar to plasma exposures, extrapolated urinary concentrations of dapagliflozin in mouse, rat, and dog toxicity studies were >700× relative to humans, and extrapolated urinary concentrations of the D3OG metabolite were 1–15× in rodents and 30× in dogs relative to humans.

Off-Target Pharmacology: Dapagliflozin and its primary human metabolite, D3OG were screened in more than 300 secondary pharmacology assays that include enzyme inhibition and receptor binding for potential off target activity. No significant off-target pharmacology was observed at pharmacologically relevant concentrations in any of these in vitro assays, suggesting that dapagliflozin and dapagliflozin-3-O-glucuronide do not exhibit off-target pharmacology.

Cell Proliferation: Since increased cellular proliferation may be associated with tumor promotion, the incidence of urinary bladder hyperplasia was assessed in the mouse and rat carcinogenicity studies and in the chronic dog toxicity studies. Dapagliflozin did not induce urinary bladder tumors in either the 2-year mouse or rat carcinogenicity studies at systemic exposures that were >70× human exposures at the maximum recommended human dose (MRHD) [37]. Dapagliflozin did not increase the incidence of hyperplasia in the urinary bladder in either the 2-year mouse or rat carcinogenicity studies. There was also no evidence that dapagliflozin directly increased the incidence of urinary bladder hyperplasia following daily administration for up to 12-months in dogs at systemic exposures that were >3000× exposure in humans at the MRHD. In vitro assessments indicate that dapagliflozin and 3ODG did not enhance cell proliferation when tested in six human bladder transitional cell lines [37]. Additionally, gene transcription analysis conducted on rat pharmacology studies with dapagliflozin indicated that dapagliflozin administration had no effect on cell cycling, cell regulation, or cell contact gene expression in the liver, kidney, skeletal muscle, or adipose tissues [37].

SGLT2 Inhibition and Increases in Urinary Glucose: A series of experiments was conducted to test whether SGLT2 inhibition and resulting increase in urinary glucose could be associated with the tumor promotion or progression. SGLT2 knockout (KO) and wild type (WT) mice were maintained until they were 15 months of age. Despite a lifetime of glucosuria, 86 % of the KO mice survived compared to 85 % of the WT mice. There was also no evidence of any renal dysfunction in the KO mice. Microscopic evaluation of the urinary bladder, kidneys, liver, heart, pancreas, adrenal glands, thyroids, spleen, female reproductive tract, male sex glands, skin, brain, and skull did not reveal any adverse effect attributable to prolonged exposure to glucosuria. Of particular note, no hyperplasia or neoplasia was observed in the urinary bladder mucosa, urogenital tract, or kidneys of SGLT2 KO mice compared to WT controls. It should also be noted that increases in urinary glucose (up to 400–500 mM) was a common feature of the toxicity studies conducted with dapagliflozin [37]. In no case were increases in urinary glucose associated with urinary bladder hyperplasia or the development of tumors. To test for a potential association between increasing glucose concentrations and transitional cell bladder tumors, an in vitro experiment was conducted in which six human bladder transitional cell lines were exposed to increasing concentrations of glucose in the media. The growth of all cell lines was completely inhibited at 50 mM glucose, well below the concentrations of glucose measured in clinical studies with dapagliflozin (mean of 166 mM at 10 mg dose; [28]). Based on this work, it was concluded that increases in urinary glucose excretion observed in the clinic were unlikely to contribute to enhanced growth of bladder tumors in patients.

Tumor Progression: To assess dapagliflozin and tumor progression, dapagliflozin was administered to nude mice implanted with either EJ-1 or UM-UC-3 (human transitional cell carcinoma cell lines) tumor xenografts. Dapagliflozin was not associated with growth of either tumor implants at doses 75× human clinical exposures [37].

Dapagliflozin was approved in the EU on November 2012 and in the US on January 2014. As part of the approval in the US, the FDA required a nonclinical post-marketing requirement. The FDA required the sponsor to evaluate dapagliflozin in an orthotopic rodent bladder tumor promotion model.

10.3.2 Prasugrel

Prasugrel (Ef(f)ient™),a novel member of the thienopyridine class of antiplatelet agents that includes ticlopidine and clopidogrel, is indicated for the reduction of thrombotic cardiovascular events (including stent thrombosis) in patients with acute coronary syndrome. Prasugrel is an orally administered prodrug requiring in vivo metabolism to form an active metabolite (R-138727) which specifically and irreversibly antagonizes the $P2Y_{12}$ class of adenosine diphosphate (ADP) receptors on platelets, thus inhibiting ADP- mediated platelet activation and aggregation. Following oral administration, prasugrel is rapidly metabolized to form the active metabolite (R-138727) which is further metabolized and inactivated to R-106583, the major circulating metabolite in humans. The metabolic pathways of prasugrel in mice, rats, and humans are generally similar.

A comprehensive nonclinical safety assessment including genotoxicity and carcinogenicity studies supported the chronic use of prasugrel in patients with atherothrombotic disease [13, 49]. Prasugrel was negative in a battery of genotoxicity studies: the Ames bacterial reverse mutation assay; an in vitro chromosomal aberration study in Chinese Hamster Lung cells; and an in vivo mouse bone marrow micronucleus test. When tested in traditional chronic rodent bioassays at exposures up to 74-fold (inactive metabolite) to 1081-fold (active metabolite) higher than the clinical exposure during 10-mg/day maintenance dosing, prasugrel was negative in a 2-year rat carcinogenicity study. Statistically significant increases in hepatocellular adenomas were seen in the 2-year mouse carcinogenicity study at a dose 250-fold the clinical exposure. The increase in liver tumors in prasugrel-treated mice was considered related to hepatic drug-metabolizing enzyme induction and not of significant relevance to human risk [4]. There were no increases (statistically significant compared to concurrent controls or compared with historical control data) in tumor incidence for either species at any site except for the increased incidence of tumors in the mouse liver.

However, in the Phase 3 clinical trial with prasugrel (and a comparator antiplatelet agent, clopidogrel), excess neoplasms (all types) were reported in the prasugrel group: the frequency of newly diagnosed cancers was 1.6 % in the prasugrel group versus 1.2 % in the clopidogrel group [46]. It must be noted that the Phase 3 study was not designed to capture baseline cancer information (e.g., cancer history, stage at diagnosis, and treatment), and the tumor observations were collected "ad hoc" through adverse event reporting. Due to regulatory concerns about the apparent increase in malignancies, additional nonclinical studies were undertaken to examine the possibility that prasugrel may accelerate tumor growth [4]. These studies included both in vitro evaluations of human tumor cell lines in culture and in vivo nude mouse xenograft models of lung, colon, and prostate origin, collectively allowing an evaluation of diverse human tissue types that may be relevant to human disease.

<u>Tumor cell proliferation</u> (<u>in vitro</u>): Prasugrel's active and inactive metabolites did not increase tumor cell proliferation in human lung, colon, or prostate tumor cells in vitro. The data also demonstrated that the assay conditions employed in these studies maintained the ability of the cells to respond to mitogenic stimuli as shown by the response to fetal bovine serum.

<u>Tumor progression</u> (<u>in vivo</u>): In the in vivo tumor xenograft studies, exposures (AUC) at 10-mg/kg doses in the tumor-bearing nude mice were approximately 34-fold higher than the exposure to R-138727 and 22-fold higher for the exposure

to R-106583 in humans administered 10-mg maintenance doses. There was no significant difference in mean terminal body weights for control versus treated animals for each tumor type. Prasugrel did not increase tumor cell proliferation (tumor volumes or weights and tumor growth rate) in human colon, lung, or prostate tumor cells in vivo.

Pharmacology: Prasugrel is highly specific for its target, the $P2Y_{12}$ class of adenosine diphosphate (ADP) receptors on platelets. Consistent with the experimental findings, there is little plausible biological basis for antithrombotic agents such as prasugrel to be carcinogenic; research has, in fact, suggested the opposite (described in [4]). It is generally accepted that prohemostatic or prothrombotic pathways, namely the coagulation cascade and platelet activation and aggregation are pro-carcinogenic. Several reports have concluded that the co-aggregation of platelets with tumor cells provide a means for tumor cells to travel to distal sites and metastasize and to avoid immune surveillance. Accordingly, preclinical studies have documented the tumor-inhibitory activity of both anticoagulants and antiplatelet agents (as referenced in [4]).

Ascertainment bias was suspected, e.g., that the higher incidence of bleeding in the prasugrel (vs clopidogrel) arm of the Phase 3 clinical study resulted in additional medical attention, during which more pre-existing cancers were discovered. In addition, several factors were weighed in the FDA's consideration as to whether prasugrel was causally related to the higher rate of tumors in prasugel treated patients in the Phase 3 study [46]:

- It was difficult to conceptualize a potential mechanism through which prasugrel could initiate or stimulate nonspecific tumor development.
- Given the relatively brief duration of the study (15 months) and the early emergence of many of the tumors, it was not thought that induction of new tumors could plausibly explain the increase.
- In silico structure activity assessment suggested that prasugrel is not carcinogenic. There were no proliferative signals (e.g., hyperplasia) in the rodent carcinogenicity studies or in chronic studies in rats or dogs. Moreover, animal carcinogenicity studies of prasugrel were negative (with the exception of the clinically irrelevant mouse liver tumors).
- Prasugrel was negative in tumor-progression studies to assess the potential effects of prasugrel and its metabolites in human colon, lung, and prostate tumor-cell lines grown in vitro and in congenitally immunodeficient "nude" mice in vivo. To FDA knowledge, the only products thought to stimulate tumor development are the erythropoietins, which, unlike prasugrel, are growth factors.
- Finally, given the observational nature of safety analyses, the fact that numerous comparisons were performed without statistical correction, and the lack of prespecified hypotheses, as well as the marginal statistical support for the finding, the possibility of a false positive finding seemed high.

The FDA and a Scientific Advisory Panel concluded that causality between prasugrel treatment and tumorigenicity or tumor promotion was unlikely. The Sponsors were assigned a postmarketing requirement to collect baseline and subsequent data on cancer in a large, at-the-time ongoing clinical trial.

The results of another clinical investigation of Dual Antiplatelet Therapy (DAPT; [32]) highlight the complexity of analyzing a potential tumor signal in clinical trials. In the DAPT study, subjects received dual antiplatelet therapy (either clopidogrel or prasugrel) beyond 1 year in duration. This study clearly showed a reduction in both stent thrombosis and myocardial infarction when dual antiplatelet therapy is extended beyond 1 year after implantation of a drug-eluting stent; however, there was an observed increase in moderate or severe bleeding, as well as a possible increase in all-cause mortality. While the study might be considered of sufficient duration (12–33 months) to test for some treatment-related signal of cancer, limitations in the study design (e.g., inconsistent reporting of cancer and characterization of cancer history) rendered any relationship of rates of cancer deaths per treatment to study drug uncertain. Additional blinded adjudication initially revealed a statistical increase in cancer-related deaths; however, the apparent increase was subsequently determined to be related to an imbalance in patient entry criteria. The added adjudication process discovered there were patients who had entered the study with advanced cancer, and there was an imbalance at baseline of eight vs one in the two respective groups of 30- versus 12- months' thienopyridine treatment; when these patients are removed, the non-cardiovascular deaths were no longer statistically significant. This initial finding prompted a meta-analysis of more than 69,000 clopidogrel-treated patients with over 139,000 patient years which showed that extended duration dual antiplatelet therapy was not associated with a difference in the risk of all-cause, cardiovascular, or non-cardiovascular death compared with aspirin alone or short duration dual antiplatelet therapy [12]. Analyses of the DAPT study highlight the importance of duration (sufficient to examine treatment-emergent development of cancer), sufficient experience (the meta-analysis allowed assessment of significant numbers of patients), and understanding bias (in this case, enrollment bias) when assessing potential cancer signals in clinical trials.

10.3.3 Cladribrine

Clabridine (Litak) was approved in the EU in 2004 [17] for the treatment of hairy cell leukemia (HCL). It is an antimetabolite chemically derived from deoxyadenosine, where the hydrogen atom in the two-position of the purine ring has been replaced by a chlorine atom, thus rendering the molecule resistant to the deamination by adenosine deaminase. Intracellularly, clabridine is phosphorylated by deoxycytidine kinase (which is present in a high concentration particularly in normal and malignant lymphoid cells). Because lymphoid cells also have a low content of 5′-nucleotidase, there is accumulation of two- chlorodeoxyadenosine-5′-triphosphate (CdATP) which is incorporated into DNA strands, thereby blocking DNA chain elongation, inhibiting DNA repair and ribonucleotide reductase. Cell death then occurs from energy depletion and apoptosis.

Clabridine is a cytotoxic medicinal product shown to be mutagenic to cultured mammalian cells. In vitro studies in various cell lines have shown that clabridine induces dNTP imbalance, DNA strand breaks, depletion of NAD and ATP, and cell death. It also inhibits DNA repair. These properties support generally the proposed mechanism of action and the therapeutic effect of clabridine. The carcinogenic potential of clabridine was tested in a single 22-month study in mice and in a TgrasH2 transgenic mouse bioassay. In the 22-month study, a significant increase in Harderian gland tumors was observed. Except for three adenocarcinomas in the high dose group, tumors were mostly benign adenomas, and there were no histomorphologic signs of progression to adenocarcinomas. Harderian gland tumors were not considered clinically relevant, as humans do not have a comparable anatomical structure [6]. The TgrasH2 transgenic mouse bioassay was negative, and the absence of any sign of Harderian gland alteration was considered to further add to the conclusion regarding the clinical irrelevance of Harderian gland tumors. The EMA Committee for Medicinal Products for Human Use (CHMP) concluded that, overall, the mouse studies did not reveal evidence of clinically relevant carcinogenic potential of clabridine [18, 19].

In HCL patients, there appeared to be no evidence that clabridine-treated patients had a higher frequency of secondary malignancies than patients treated with alpha-interferon or deoxycoformycin. However, since the incidence of secondary malignancies was significantly higher compared to the general population, the CHMP recommended that patients treated with clabridine be regularly monitored and that an annual follow-up report on secondary malignancies be provided [18, 19]. Warnings about secondary malignancies and regular monitoring as a precaution were incorporated in the SPC.

In 2009, an oral tablet formulation of clabridine (Movectro) was developed for the treatment of Multiple Sclerosis (MS). As of August 2010, five malignancies in clabridine-treated patients vs. one in placebo patients were reported. Over the entire clinical program, 22 cases of malignancies were reported in clabridine-treated MS patients, while only two cases were reported in placebo-treated patients (one basal cell carcinoma and one ovarian cancer). The Relative Risk (RR) of malignancies based on patients from all studies suggested a five-fold increase in the risk of cancer but with a broad CI (95 % CI: 0.67–38.43). However, the sponsor considered that a more appropriate estimation of RR should be derived from analyses restricted to clabridine-treated MS patients in double-blind controlled trials, thus avoiding confounding by dissimilar follow-up periods of the treated and placebo cohorts. That analysis yielded an RR of 2.31 (95 % CI: 0.27–19.81), suggesting only a two-fold increase in the risk of cancer among clabridine exposed patients which was statistically not significant. The CHMP agreed that, while RR calculation based on all studies might be biased, the more conservative analysis based on all studies was more appropriate and suggested an increased risk of malignancy with increased exposure time. The concern for the disproportion of number of malignancies in the clabridine groups compared to placebo during the whole clinical trial program contributed to the negative opinion issued by CHMP in 2010, which was reiterated after a reexamination in 2011 [18, 19]. In contrast to HCL indication, the benefit/risk for MS patients was considered negative.

The Movectro case is a good example of a situation where a potential cancer signal emerged through clinical experience in the absence of a relevant positive rodent carcinogenicity study. It must be said, however, that concern for possible human carcinogenic risk was prudent based on the positive genotoxicity and the pharmacological properties of clabridine, even though the mouse carcinogenicity bioassays results had been reassuring.

10.3.4 Avandia

Rosiglitazone (Avandia™) was approved to improve glycemic control in patients with Type 2 diabetes mellitus [14, 47]. Thiazolidinediones such as rosiglitazone produce their effects by activating peroxisomal proliferator-activated receptor gamma (PPARγ), altering gene expression associated with multiple molecular and cellular processes. The marketing application was supported by a comprehensive nonclinical safety assessment which included genotoxicity and carcinogenicity studies [47]. Genotoxicity tests of chromosomal aberration, unscheduled DNA synthesis, and the in vivo mouse micronucleus were negative, while the incidence of forward mutations at the TK locus of mouse lymphoma L5178Y cells was slightly increased (ca. 2×) in the presence of S-9. In the rodent carcinogenicity studies, there were no remarkable findings except an increase in the incidence of adipose hyperplasia in mice and significant increases in benign adipose tissue tumors (lipomas) in rats. The proliferative changes in both species were considered due to the persistent pharmacological overstimulation of adipose tissue.

Rosiglitazone is an example of a drug for which there was no signal of animal (excepting the target-related lipomas) or human tumorigenicity but which faced questions because, with piogliatozone, a related thiazolidinedione with PPARγ activity, tumors were observed in the urinary bladder of male rats in a 2-year carcinogenicity study [45] and there were reports of bladder cancer in some patients taking the drug [44]. Clinically, rosiglitazone has received little attention regarding bladder cancer risk in patients, in large part due to FDA-imposed stringent prescribing restrictions and the EMA suspension of the marketing authorization related to cardiovascular risks [15, 48]. In 2004, a number of PPARγ agonists were being screened for potential chemopreventive properties in a nonclinical model. Lubet et al. [29] reported that a relatively high dose of rosiglitazone appeared to promote bladder cancer formation in the hydroxybutyl(butyl)nitrosamine (OHBBN, a urinary bladder specific carcinogen) -induced rat bladder tumor model. At that time, the FDA was reporting that a number of recently synthesized PPARγ and PPARα/γ agonists were themselves inducing bladder tumors in rats or mice or in both species [11]. A subsequent 2-stage tumor promotion study expanded upon the initial data, suggesting that lower doses of rosiglitazone may also have significant tumor promoting activity in the OHBBN rat model [30]. The potential for rosiglitazone to be associated with later-stage promotional activity was considered somewhat surprising in that the PPARγ receptor, although highly expressed in normal bladder urothelium and hyperplastic lesions, was expressed at lower levels in established bladder cancers.

This case study provides an example in which a rat model of initiation and promotion was employed to investigate an initial report of potential tumor promoting activity of rosiglitazone (in that same model) and a concern regarding the pharmacologic class [11]. However, unlike pioglitazone, there have been no reports of any theoretical association of rosiglitazone treatment and bladder hyperplasia or cancer in animals or patients. In a systematic review and meta-analysis of clinical trial and observational studies, Turner et al. [44] concluded that "no significant risk was seen with rosiglitazone" and that "the evidence for any relationship between bladder cancer risk and rosiglitazone cumulative duration is limited and inconsistent."

10.3.5 Orlistat

Orlistat is a specific and long-acting inhibitor of pancreatic and gastric lipases and is currently marketed as a treatment for obesity in the US and Europe as both a prescription (Xenical™, Roche) and over-the-counter (Alli, GlaxoSmithKline) drug. Orlistat is a partially hydrated derivative of lipstatin that functions by decreasing the breakdown and subsequent absorption of an estimated one-third of dietary ingested fats [33]. Pharmacokinetic studies indicate that orlistat has very low oral bioavailability and suggest that the effects of orlistat are restricted to the intestines [54]. In a 2-year efficacy study, obese patients receiving 120 mg orlistat three times a day lost significantly more weight (8.8 %) than those patients receiving placebo (5.8 %) after the first year of the study. During the second year, twice as many patients receiving placebo (63 %) regained their weight compared to those maintained on orlistat (35 %). The most common adverse events observed in patients receiving orlistat included abdominal pain, fatty/oily evacuation, and fecal incontinence [33].

During the Phase 3 clinical trials, nine cases of breast cancer were observed in women taking orlistat compared to one patient in the placebo group. During follow-up surveys, two more patients receiving orlistat (11 total) were diagnosed with breast cancer compared to three in the placebo group [31]. The FDA indicated that the data submitted supported the efficacy of orlistat but asked Roche to gather further information on the breast cancer cases observed in the clinical trials. The reason for the clinical imbalance in breast cancer was unknown but was speculated to be due to chance or detection bias. In August 1997, Roche withdrew its NDA and then resubmitted it in November 1997 [31]. At the Xenical™ FDA Advisory Committee meeting (March 1998), independent experts in the fields of oncology, histopathology, and mammography agreed that the majority of the breast cancers observed in the orlistat clinical trials were pre-existing and that 3 of the cases in the orlistat treatment group and two in the placebo group emerged after treatment initiation [38]. Therefore, with this new data, there was no difference in the incidence of breast cancer in patients treated with orlistat compared to placebo [33]. Nonclinical studies were also supportive of a lack of a tumor risk with orlistat [38]. Orlistat did not induce tumors in the 2-year rodent carcinogenicity studies and was not genotoxic in nonclinical testing. Therefore, it was the opinion of the experts at the Advisory Committee meeting that there was no evidence that orlistat induced breast cancer. More data from open-label Phase 3b trials confirmed this conclusion. In these trials, three additional cases of breast cancer were observed (all in the placebo group) with no observed imbalance in breast cancer cases [50]. This additional data was provided to regulatory authorities in January 1999 and lead to the approval of orlistat (Xenical™, Roche) by the FDA in April of 1999.

Interestingly, orlistat has been shown in recent years to exhibit potent antitumor activity in vitro through its ability to block cellular fatty acid synthesis activity and induce apoptosis in colon and breast cancer cells [26, 34]. Although the low bioavailability of orlistat may prevent its utility in treating breast cancer, it has been suggested that the antitumor properties of orlistat may have beneficial effects for the treatment of tumors of the gastrointestinal tract [35].

10.4 Summary

In summary, any signal of potential treatment-related malignancy should be considered and appropriately evaluated. The case studies presented illustrate the various kinds of hypothesis-driven nonclinical investigations that may be conducted to evaluate potential cancer risk when a human tumor signal is identified. While these post-hoc or retrospective assessments should take into account the biology and pharmacology of the molecule, they may also include studies addressing tumor promotion and progression on a case-by-case basis. However, it is very important to recognize that these models are not standard or well-validated, and the development and validation of innovative models for assessing tumor promotion and progression that are more human-based warrants further scientific investigation.

In practice, risk evaluation and management of potential safety signals in the settings of late-phase clinical development and/or real world use may take several forms. Risk Management Plans (RMPs; [20]) include a set of pharmacovigilance activities and interventions designed to identify, characterize and manage risks related to a medicine. Pharmacovigilance actions to investigate specific safety concerns such as tumorigenicity may include targeted safety studies, postmarketing surveillance, observational and epidemiologic studies, and mechanistic or descriptive studies. Risk minimization activities may include treatment axations, patient restriction or exclusion and updated labeling requirements.

References

1. Alden CL, Lynn A, Bourdeau A, Morton D, Sistare FD, Kadambi VJ, Sliverman L (2011) A critical review of the effectiveness of rodent pharmaceuitical carcinogenesis testing in predicting for human risk. Vet Pathol 48:772–784

2. Barret JC (1995) Mechanisms for species differences in receptor-mediated carcinogenesis. Mutat Res 333(1–2):189–202
3. Bhatia SN, Ingber DE (2014) Microfluidic organs-on-chips. Nat Biotechnol 32:760–772
4. Buckley LA, Sanbuissho A, Starling JJ, Knadler MP, Iversen PW, Jakubowski JA (2012) Nonclinical assessment of carcinogenic risk and tumor growth enhancement potential of prasugrel, a platelet-inhibiting therapeutic agent. Int J Toxicol 31(4):317–325
5. Bugelski PJ, Volk A, Walker MR, Krayer JH, Martin P, Descotes J (2010) Critical review of preclinical approaches to evaluate the potential of immunosuppressive drugs to influence human neoplasia. Int J Toxicol 29:435–466
6. Carlton WW, Render JA (1991) Adenoma and adenocarcinoma in harderian gland, mouse rat and hamster. In: Jones TC, Mohr U, Hunt RD (eds) Monographs on pathology of laboratory animals, eye and ear. Springer, Heidelberg, pp 133–137
7. Chen J, Williams S, Ho S, Loraine H, Hagan D, Whaley JM, Feder JN (2010) Quantitative PCR tissue expression profiling of the human SGLT2 gene and related family members. Diabetes Ther 1(2):57–92
8. Chen TX, Wanibuchi H, Murai T, Kitano M, Yamamoto S, Fukushima S (1999) Promotion by sodium L-ascorbate in rat two-stage urinary bladder carcinogenesis is dependent on the interval of administration. Jpn J Cancer Res 90(1):16–22
9. Cohen SM (1998) Cell proliferation in the evaluation of carcinogenic risk and the inadequacies of the initation-promotion model. Int J Toxicol 17(Suppl 3):129–142
10. Dorato MA, McMillian CL, Vodicnik MJ (2008) The toxicology assessment of pharmaceutical and biotechnology products. In: Hayes AW (ed) Principles and methods of toxicology, 5th edn. CRC Press/Taylor and Francis Group, Boca Raton, FL pp 325–367
11. El-Hage J (2005) Peroxisome proliferation-activated receptor agonists: carcinogenicity findings and regulatory recommendations. In: International Atherosclerosis Society Symposium on PPAR. Monte Carlo
12. Elmariah S, Mauri L, Doros G, Galper BZ, O'Neill KE, Steg PG, Kereiakes DJ, Yeh RW (2014) Extended duration dual antiplatelet therapy and mortality: a systematic review and meta-analysis. Lancet. http://dx.doi.org/10.1016/S0140-6736(14)62052-3
13. European Medicines Agency (EMA) European Public Assessment Report (EPAR) of Efient (2009) EMEA/H/C/984. http://www.ema.europa.eu/docs/en_GB/document_library/EPAR_-_Summary_for_the_public/human/000984/WC500021973.pdf. Accessed 04 Nov 2014
14. European Medicines Agency (EMA) Scientific Discussion for Approval of Avandia (2003) http://www.ema.europa.eu/docs/en_GB/document_library/EPAR_-_Scientific_Discussion/human/000268/WC500029103.pdf. Accessed 22 Dec 2014
15. European Medicines Agency (EMA) Assessment Report for Avandia (2010) EMA/831385/2010. http://www.ema.europa.eu/docs/en_GB/document_library/EPAR_-_Assessment_Report_-_Variation/human/000268/WC500100757.pdf
16. European Medicines Agency (EMA) Assessment Report on Calcitonin-Containing Medicinal Products (2013) EMA/109665/2013. http://www.ema.europa.eu/docs/en_GB/document_library/Referrals_document/Calcitonin_31/WC500146172.pdf. Accessed 04 Nov 2014
17. European Medicines Agency (EMA) Scientific Discussion of Litak (Cladribine) (2005) http://www.ema.europa.eu/docs/en_GB/document_library/EPAR_-_Scientific_Discussion/human/000504/WC500041660.pdf. Accessed 3 Nov 2014
18. European Medicines Agency (EMA) (2011) Withdrawal Assessment Report for Movectro. http://www.ema.europa.eu/docs/en_GB/document_library/Application_withdrawal_assessment_report/2011/03/WC500104393.pdf. Accessed 03 Nov 2014
19. European Medicines Agency (EMA) Withdrawal Assessment Report for Movectro (2011) http://www.ema.europa.eu/docs/en_GB/document_library/Application_withdrawal_assessment_report/2011/03/WC500104393.pdf. Assessed 3 Nov 2014
20. European Medicines Agency (EMA) Guideline on good pharmacovigilance practices (GVP) (2014) EMA/838713/2011 Rev 1*. http://www.ema.europa.eu/docs/en_GB/document_library/Scientific_guideline/2012/06/WC500129134.pdf. Accessed 06 Oct 2014

21. Friedrich A, Olejniczak K (2011) Evaluation of carcinogenicity studies of medicinal products for human use authorised via the European centralised procedure (1995–2009). Regul Toxicol Pharmacol 60(2):225–248
22. Huff JE (1994) Chemicals causally associated with cancers in humans and in laboratory animals: a perfect concordance. In: Waalkes MP, Ward JM (eds) Carcinogenesis. Raven, New York, pp 25–37
23. Jacobs EJ, Henion AK, Briggs PJ, Connell CJ, McCullough ML, Jonas CR, Rodriguez C, Calle EE, Thun MJ (2002) Vitamin C and vitamin E supplement use and bladder cancer mortality in a large cohort of US men and women. Am J Epidemiol 156(11):1002–1010
24. Johnsson K, Ptaszynska A, Apanovitch AM, Sugg J, Parikh S, List J (2012) Safety of dapagliflozin in clinical trials for T2DM. Presented at: European Association for the Study of Diabetes. Berlin; (Abstract 743)
25. Kanai Y, Lee WS, You G, Brown D, Hediger MA (1994) The human kidney low affinity Na+/glucose cotransporter SGLT2. Delineation of the major renal reabsorptive mechanism for D-glucose. J Clin Invest 93(1):397–404
26. Kridel SJ, Axelrod F, Rozenkrantz N, Smith JW (2004) Orlistat is a novel inhibitor of fatty acid synthase with antitumor activity. Cancer Res 64(6):2070–2075
27. Link WT, De Felice A (2014) An FDA overview of rodent carcinogenicity studies of angiotensin II AT-1 receptor blockers: pulmonary adenomas and carcinomas. Regul Toxicol Pharmacol 70:555–563
28. List JF, Woo V, Morales E, Tang W, Fiedorek FT (2009) Sodium-glucose cotransport inhibition with dapagliflozin in type 2 diabetes. Diabetes Care 32(4):650–657
29. Lubet RA, Steele VE, Crowell JA, Hara Y, Juliana MM, Eto I, Grubbs CJ (2004) Effects of celecoxib, polyphenon E and rosiglitazone on hydroxybutyl(butly)-nitrosamine (OH-BBN)-induced urinary bladder cancers in female Fischer-344 rats. Prevention Res 5: Chemoprevention Studies II. Abstract 3132
30. Lubet RA, Fischer SM, Steele VE, Juliana MM, Desmond R, Grubbs CJ (2008) Rosiglitazone, a PPAR gamma agonist: potent promoter of hydroxybutyl(butyl)nitrosamine-induced urinary bladder cancers. Int J Cancer 123:2254–2259
31. Malone M, Orlistat F (1998) Hoffmann-La Roche Ltd. Drugs 1(2):232–241
32. Mauri L, Kereiakes DJ, Yeh RW et al (2014) Twelve or 30 months of dual antiplatelet therapy after drug-eluting stents. New Engl J Med. doi:10.1056/NEJMoa1409312
33. McNeely W, Benfield P (1998) Orlistat. Drugs 56(2):241–249
34. Menendez JA, Vellon L, Lupu R (2005) Antitumoral actions of the anti-obesity drug orlistat (XenicalTM) in breast cancer cells: blockade of cell cycle progression, promotion of apoptotic cell death and PEA3-mediated transcriptional repression of Her2/neu (erbB-2) oncogene. Ann Oncol 16(8):1253–1267
35. Menendez JA, Vellon L, Lupu R (2005) Orlistat: from antiobesity drug to anticancer agent in Her-2/neu (erbB-2)-overexpressing gastrointestinal tumors? Exp Biol Med (Maywood) 230(3):151–154
36. Obermeier M, Yao M, Khanna A, Koplowitz B, Zhu M, Li W, Komoroski B, Kasichayanula S, Discenza L, Washburn W, Meng W, Ellsworth BA, Whaley JM, Humphreys WG (2010) In vitro characterization and pharmacokinetics of dapagliflozin (BMS-512148), a potent sodium-glucose cotransporter type II inhibitor, in animals and humans. Drug Metab Dispos 38(3):405–414
37. Reilly TP, Graziano MJ, Janovitz EB, Dorr TE, Fairchild C, Lee F, Chen J, Wong T, Whaley JM, Tirmenstein M (2014) Carcinogenicity risk assessment supports the chronic safety of dapagliflozin, an inhibitor of sodium-glucose co-transporter 2, in the treatment of type 2 diabetes mellitus. Diabetes Ther 5(1):73–96
38. Roche 1998 (1988) Xenical® (orlistat) Advisory Committee briefing document. Presented at the Advisory Committee
39. Sabolic I, Vrhovac I, Eror DB, Gerasimova M, Rose M, Breljak D, Ljubojevic M, Brzica H, Sebastiani A, Thal SC, Sauvant C, Kipp H, Vallon V, Koepsell H (2012) Expression of Na+-D-glucose cotransporter SGLT2 in rodents is kidney-specific and exhibits sex and species differences. Am J Physiol Cell Physiol 302(8):C1174–C1188

40. Silva Lima B, van der Laan JW (2000) Mechanisms of nongenotoxic carcinogenesis and assessment of the human hazard. Regul Toxicol Pharmacol 32(2):135–143
41. Swann JB, Smyth MJ (2007) Immune surveillance of tumors. J Clin Invest 117(5):1137–1146
42. Thomas S, Chiriva-Internati M, Shah GV (2007) Calcitonin receptor-stimulated migration of prostate cancer cells is mediated by urokinase receptor-integrin signaling. Clin Exp Metastasis 24:363–377
43. Tirmenstein M, Dorr TE, Janovitz EB, Hagan D, Abell LM, Onorato JM, Whaley JM, Graziano MJ, Reilly TP (2013) Nonclinical toxicology assessments support the chronic safety of dapagliflozin, a first-in-class sodium-glucose cotransporter 2 inhibitor. Int J Toxicol 32(5): 336–350
44. Turner RM, Kwok CS, Chen-Turner C (2013) Thiazolidinediones and associated risk of bladder cancer: a systematic review and meta-analysis. Br J Clin Pharmacol 78(2):258–273
45. Tseng C-H, Tseng F-H (2012) Peroxisome proliferatory-activated receptor agonists and bladder cancer: lessons from animal studies. J Environ Sci Health C Environ Carcinog Ecotoxicol Rev 30:368–402
46. Unger EF (2009) Weighing benefits and risks–the FDA's review of prasugrel. N Engl J Med 361(10):942–945
47. US Food and Drug Administration (USFDA) Drug Approval Package Avandia (NDA 21–071) (1999) http://www.accessdata.fda.gov/drugsatfda_docs/nda/99/21071_Avandia_Approv.pdf. Accessed 18 Dec 2014
48. US Food and Drug Administration (USFDA) (2011) FDA drug safety communication: updated Risk Evaluation and Mitigation Strategy (REMS) to restrict access to rosiglitazone-containing medicines including Avandia, Avandamet, and Avandaryl. http://www.fda.gov/Drugs/DrugSafety/ucm255005.htm. Accessed 22 Dec 2014
49. US Food and Drug Administration (USFDA). EFFIENT™ Summary Review (NDA 22–307) (2009) http://www.accessdata.fda.gov/drugsatfda_docs/nda/99/021073A_Actos_phrmr_P1.pdf. Accessed 04 Nov 2014
50. Van R, Walt L (1999) Outdated and misleading review of orlistat. Can Fam Physician 45:2849–2850, 2852
51. Vyas U (2012) Risk of breast cancer due to hyperprolactinemia caused by antipsychotics (neuroleptics). Brit J Med Pratict 5(4):a534–a538
52. Williams GM, Iatropoulos MJ, Enzmann HG (2008) Principles of testing for carcinogenic activity. In: Hayes AW (ed) Principles and methods of toxicology, 5th edn. CRC Press/Taylor and Francis Group, Boca Raton, FL pp 1265–1316
53. Wookey PJ (2009) A review of calcitonin receptor expression in embryonic, foetal and adult tissues, with an hypothesis on the connection between expression during foetal development and disease. Open Zool J 2:53–61
54. Zhi J, Melia AT, Eggers H, Joly R, Patel IH (1995) Review of limited systemic absorption of orlistat, a lipase inhibitor, in healthy human volunteers. J Clin Pharmacol 35(11):1103–1108

Index

M.J. Graziano, D. Jacobson-Kram (eds.), *Genotoxicity and Carcinogenicity Testing of Pharmaceuticals*, DOI 10.1007/978-3-319-22084-0

Printed by Printforce, the Netherlands